BIG BANG, BLACK

DAVID TOBACK

MW00844255

"I THINK YOU SHOULD BE MORE
EXPLICIT HERE IN STEP TWO."

Foreword by Eiichiro Komatsu

Kendall Hunt
p u b l i s h i n g   c o m p a n y

Cover image Reprinted by permission. ScienceCartoonsPlus.com.

Kendall Hunt
publishing company

www.kendallhunt.com
Send all inquiries to:
4050 Westmark Drive
Dubuque, IA 52004-1840

ISBN 978-1-4652-2578-8

Printed in the United States of America
10 9 8 7 6 5 4 3 2 1

Contents

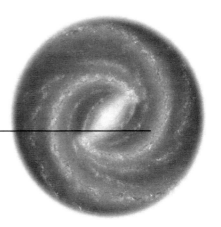

Foreword

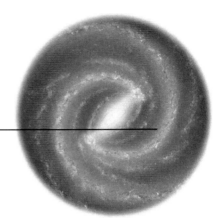

This book has a clear and singular goal: to explain recent exciting developments in cosmology—our understanding of how the universe works—in everyday language.

This book grew out of Prof. Toback's lectures on cosmology for non-science majors at Texas A&M University. I wish this book were published when I was teaching cosmology in my class! I agree with everything Prof. Toback says in the Preface. Our knowledge of the universe has been truly revolutionized over the last twenty years or so. However, for some reason, this revolution has not been communicated well to the people outside of the science community. The knowledge that a vast majority of people would have on the universe is most likely obsolete, as we have new and far more accurate knowledge about the universe now than twenty years ago. I am a professional cosmologist, and cosmology is what I do for living. I have also tried to teach cosmology for non-science majors several times over the last nine years, but frankly, I do not think it was successful. A part of the difficulty was, as Prof. Toback also points out in Preface, the lack of suitable textbooks.

Big Bang, Black Holes, No Math fills the gap. This is exactly the kind of book needed to communicate the exciting, revolutionizing developments in our understanding of the universe to the general public, in the language that anyone can understand. This book is full of useful analogies and illustrations which will greatly help the readers relate complex phenomena in the universe to their daily experiences. And this book does not contain any equations, as advertised.

The overarching question is: we now think that the universe "inflated" shortly after its birth; became hot like a fireball; created hydrogen and helium by nuclear fusion in the first three minutes; and is full of dark matter and dark energy, whose nature is currently unknown. These are the essential elements of our own universe, but what do all these mean? This book will get the readers up to speed on the forefront of cosmological studies.

I hope that learning and understanding how the universe works gives a joy not only to scientists who study it for living, but also to everyone who lives in it. *Big Bang, Black Holes, No Math* provides an excellent starting point. Read this book, and immerse yourself in the vastness of the cosmos.

May 2013
Eiichiro Komatsu
Director
Max-Planck-Institute for Astrophysics
Garching, Germany

Preface

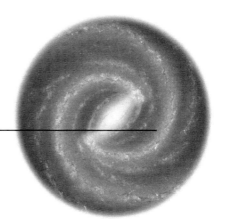

This book originated out of a desire to teach a focused class on the big bang, black holes, dark matter, and cosmology to non-science majors. I thought back to when I was in college and how much I wanted to learn about these exciting topics. While some of the subjects were covered in basic astronomy classes, they were not the emphasis of the courses, but merely afterthoughts or covered too quickly.

In physics classes, I did not get to investigate this material in-depth until I was a graduate student. Even then, the classes spent a great deal of time on advanced mathematics and didn't get to all the interesting and exciting topics intrinsic to a study of cosmology.

Also, the field of cosmology has exploded in the last ten years. It continues to be a great time of enormous progress in our understanding of the universe. Today's students want and deserve to hear about these topics front and center.

I know that many non-science students want to know more, but are apprehensive about taking physics or astronomy classes. They want to learn about the universe in a way that does not require calculations or learning about the constellations or parallax. They deserve a course tailor-made for the non-science major, yet one that does not carry the "football physics" moniker. This needs to be a very different type of course, where non-majors can undertake focused reading, discussion, and writing, delving deeply into some of the most exhilarating mysteries of our universe.

As example students, I thought of my parents when they were young. While incredibly smart and well-educated (they both have Ivy League educations and PhDs), they do not know physics. Ultimately, I wanted a course they, or any other non-cosmologist, could take. I wanted to teach the course I would have enjoyed in college.

This book has a number of goals. One is to share with you just how exciting our universe is today and how exciting it has been throughout its history. A second important goal is for you to understand some of the evidence that gives scientists confidence that the history we will tell reflects what actually happened.

One way to measure if these goals have been accomplished is if you, the reader, can communicate this understanding and enthusiasm to others. I want you to be able to effectively share it with your family, friends, co-workers, community leaders, and policymakers, and explain to them why it is exciting and important. My hope is that senators and members of the House of Representatives one day find this description (either from my book or from you directly) understandable and compelling. If you can explain these ideas to your parents and grandparents, I have done my job.

I will admit that I am generally unhappy about how science is often taught, and about how little science the general population actually understands. Many people view science as being beyond their intellectual capabilities, or just plain dull. I cannot count the number of times I have told my airplane-seat neighbors that I am a physicist, and they reply with something such as, "Ugh… I hated that course in high school. It was boring and WAY too hard." Frankly, I do not blame them.

Part of the problem is that physics and astronomy tend to be taught for *use* (for scientists- and/or engineers-in-training) rather than for enjoyment. If you are going to *use* physics or astronomy, it requires math—and a lot of it. This makes it no fun for some and a complete turn-off for many others.

Another problem is that scientists see math as the "language of the universe." Without being able to speak in their native tongue, explaining complicated scientific ideas can present a real challenge for professors. There is an uneasy sense that the full beauty is lost, or that somehow the real ideas will be inaccurately conveyed. But when this challenge is accepted and approached in the right way, a happy middle-ground can be reached. While it may not be as "satisfying" for the teacher, non-scientists reap the rewards of understanding and appreciating the beauty of our universe without the intimidation of complex calculations.

Essentially, I see similarities between science and music. Many different types of people love music, but very few are professors of music or composers. Some people play music for a living. Others just play for fun. But most just listen and enjoy it.

Music lessons and courses in music are not reserved for future composers and musicians. But for some reason, we do not approach science in this way. We should be teaching science the way music appreciation classes are taught. You do not need to play the violin or the guitar to enjoy what is on the radio. Perhaps a better title for this book should be "Cosmology Appreciation."

The course that emerged from these ideas has been a lot of fun both for my students and for me. The problem we faced, however, was the lack of a standard textbook. While there are many good books available that touch on some of the relevant topics (see the "Suggested Reading" section), they either lack the desired focus or clarity for non-majors, or they simply don't include some of the more significant or stimulating topics.

Stephen Hawking's *A Brief History of Time*, for example, was not complete enough for a course. *The Essential Cosmic Perspective*, an excellent introductory astronomy textbook, is intended to be a complete guide to the universe, rather than an investigation of its origin and evolution.

A true introductory cosmology course without math for non-majors requires a focused book. The product you see here grew from that need. My hope is that you find it as useful as my previous students have.

There are many people who played an important role in the creation of this book, and it is hard to acknowledge all of them. I will start with the students who took the course, as well as the teaching assistants who helped me develop it and encouraged me to turn it into a book. Special thanks go to Andrew Colvin, Daniel Cruz, Lucas Macri, J.P. Quinn, David Rahmani, and Justin Rowland for creating all the figures.

Thanks to Stephen Boada, Daniel Cruz, Carli Domenico, Stephen Green, Emily Hargrove, Tanner Howell, Ting Li, Chris Maguire, Kyle Page, J.P. Quinn, Jessica Smarr, Lara Speights, Gary Toback, Phyllis Toback, and Summer Wayhan for comments on the text, and to Kara Socol and Amelia Williamson-Smith for helping to turn my lecture notes into a real book. I also want to thank Jonathan Asaadi, Ziqing Hong, J.P. Quinn, and Sean Yeager for helping write and vet the after-chapter problems (available for online use) and the Coffee Station—now, unfortunately, defunct—for being my quiet hideaway to write during this process.

On the science side, I must start with my mentors to whom I am incredibly indebted: Henry Frisch, Drew Baden, and Peter McIntyre. I would also like to thank a number of people who helped me with the science, reviewed the text in part, and helped clarify its presentation:

- Richard Arnowitt (Texas A&M University)
- Mitch Begelman (University of Colorado)
- Ed Bertschinger (Massachusetts Institute of Technology)
- Peter Brown (Texas A&M University)
- James Dent (Vanderbilt University)
- Bhaskar Dutta (Texas A&M University)

- Brian Fields (University of Illinois)
- Keely Finkelstein (Texas A&M University and University of Texas)
- Steven Finkelstein (Texas A&M University and University of Texas)
- Joe Fowler (Princeton University and University of Colorado)
- Teruki Kamon (Texas A&M University)
- William Kinney (State University of New York at Buffalo)
- Eiichiro Komatsu (University of Texas and the Max Plank Institute)
- Kevin Krisciunas (Texas A&M University)
- Jason Kumar (University of Hawaii)
- Eric Linder (Lawrence Berkeley National Laboratory)
- Lucas Macri (Texas A&M University)
- Jennifer Marshall (Texas A&M University)
- Paul Padley (Rice University)
- Casey Papovich (Texas A&M University)
- Anne Pellerin (Texas A&M University)
- Jim Pivarski (Texas A&M University)
- Christopher Sirola (University of Southern Mississippi)
- Jason Steffen (Fermi National Accelerator Laboratory)
- Nicholas Suntzeff (Texas A&M University)
- Kim-Vy Tran (Texas A&M University)
- Virginia Trimble (University of California at Irvine)
- Joel Walker (Sam Houston State University)
- William Wester (Fermi National Accelerator Laboratory)
- Craig Wheeler (University of Texas)
- Josh Winn (Massachusetts Institute of Technology)
- Sylvana Yelda (University of California at Los Angeles)

There are other players who made the process a success, including the late George Mitchell and Sheridan Lorenz for their enormous support of cosmology at Texas A&M, particularly in regard to the generous gifts that established and support the George P. and Cynthia W. Mitchell Institute for Fundamental Physics and Astronomy, of which I am a proud member. I'd also like to express my appreciation to Tim Scott in the College of Science, and Lewis Ford, Ed Fry, and George Welch in the Department of Physics and Astronomy (who deserve thanks for letting me create and teach the class in the first place). Finally, I am incredibly indebted to the crucial support from Arthur J. and Wilhelmina Thaman, who funded the named professorship I now hold.

Last, but not least, I would like to thank my beloved family for the rock-solid foundation they form, as well as the inspiration they provide. This includes Justin for being the battering ram, and to my two red-headed sons, Bobby and Aaron, for their bright-eyed and relentless spirit and imagination, and for teaching me the true meaning of patience. I would like to thank my parents, Phyllis and Gary Toback, for making me the man I am today, and, in particular, my father for making me the scientist I am today.

Finally, I thank my beloved Katherine for bringing joy, happiness, and fun into my life. I hope she is proud of this book because without her, it would never have seen the light of day. I could not have asked for a better editor and companion on this journey.

May 2013
David Toback
College Station, TX
Texas A&M University
Mitchell Institute for Fundamental Physics and Astronomy

Instructor Notes

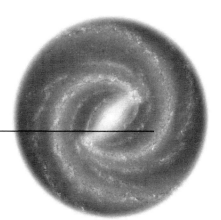

This book has been field-tested and used in a number of different types of courses and in a number of different ways. The primary use has been as the main textbook for a hundred-student Introduction to Cosmology course for non-majors. At Texas A&M University, where this course was developed, it is approved as a State of Texas Tier-2 science distribution course and is cross-listed under both astronomy and physics headings. For more information, see http://faculty.physics.tamu.edu/toback/109.

As is needed for many science distribution courses, there is a complementary laboratory methods course. More detail can be found at http://faculty.physics.tamu.edu/toback/119/. This book has also been used in a special topics course, an honors course, and a seminar course. In addition, the book has been used as a supplementary text in introductory astronomy courses for instructors who want the cosmology portion of their course to be more focused. It could easily be used for a topics course for high-school students.

It is natural to ask why this book does not look like the typical five-hundred-page, high-gloss photograph introductory textbook. It is also natural to ask why an instructor doesn't just use one of the many introduction to astronomy books already out there, which contain much of the information in this book (in some cases in a more comprehensive form). Indeed, it is certainly possible to teach an introduction to cosmology course, like a standard introductory astronomy or physics course, by choosing an exhaustive astronomy textbook, picking and choosing the reading from it, and giving exams based on the material.

The answer to these questions is that I believe a cosmology course deserves a focused cosmology book. The astronomy books on the market are wonderful, but really are designed to teach astronomy. Cosmology tends to show up as a minor topic toward the end of these books.

This book, however, concentrates on cosmology and some of the basic evidence that supports our understanding of the universe. It is worth noting that there are very few introduction to cosmology courses for non-science majors, particularly ones taught outside the context of an introductory astronomy course. Creating a cosmology course, then, is easier said than done.

By design, the chapters in this book are short and sweet and are intended to be independent of an astronomy introduction. Typically, it is about a single seventy-five-minute lecture period per chapter, although some chapters take longer than others. Since it is not designed to be all-inclusive, there is ample supplementary material for the instructor who wants to go beyond this book (online for free or in the "Suggested Reading" section). The price of this book is a fraction of the price of the typical introduction to astronomy book—a nice thing for students in a stretched economy.

There are many ways to assign homework or to create exams for a course like this. For the instructor who would like to give multiple-choice exams, there is a test bank available for this book. Instructors who have access to eLearning should feel free to request it in this format. An instructor's solutions manual is available as well.

In addition to being a standard introduction to cosmology textbook, I have written this book to be useful for the type of course that I teach: an intensive writing course. Even ten years ago, this type of course would naturally scare the typical instructor because of the enormous amount of grading it would have required in an age before online grading tools.

With today's tools, different learning-outcome objectives require powerful new evaluation methods, and many great ones exist. I happen to favor the usage of Calibrated Peer Review—"CPR" for short (see http://cpr.molsci.ucla.edu/ or http://cpr.tamu.edu/). My colleagues and I have taught this course with more than a hundred students. The time expended on our part, however, has been no greater than if it were a typical introductory course with no writing requirement.

Since it is useful to spell out an example style for the course, I'll give some details of how I do it. First of all, there are no exams, and the bulk of the course grade is based on five separate two-page papers. Each is designed to help the student attain, then demonstrate, an intuitive understanding of the central topics of the course. As such, the paper is written to be presented to a lay-reader, typically a senator or governor. The paper topics are:

1. Present the evidence that little things go around big things
2. Present the evidence for dark matter
3. Present the case that stars are made of atoms
4. Present the case that the universe started with a big bang
5. Present the case for the existence for stellar black holes

The students are graded both on the style and content of their essays. Since we are interested in teaching them how to be better evidence-based decision-makers, the goal is for the students to write these papers using their own words, and for them to convey the evidence as part of the argument.

Even in writing-intensive classes, there are other types of student work that are useful to include in the grading process. For example, about ten percent of the course grade is based on three types of activities.

The first type is a before-class writing assignment in which each student, for each chapter, must submit two questions on the reading before the lecture on that topic. They can be about things that are unclear in the reading, or if they want to know more about a topic beyond what was covered in the reading. Typically, this is accomplished online, and short responses are given to the students by the instructor or teaching assistant (if one is available), by pointing to a collected list of commonly asked questions which is posted online (and available for other instructors). Often, the responses are as simple as, "This will be discussed in a later chapter." But since the questions can be quite sophisticated, we often give more significant responses and/or point to web-links. If multiple students ask the same question, this can be used to help focus the lecture using "just in time" and "interactive engagement" teaching methods.

The second type of activity is an in-class quiz on conceptual questions. This is the modern "clicker-style" question where a question is presented to the class with a number of alternative answers. Students then must discuss in pairs and use their clicker to enter in an answer. The instructor often can interject ideas into the discussion. The results of the voting can then be further discussed. Having students raise their hand or use sheets of paper work equally well for this type of interactive-engagement activity. A set of questions is available upon request.

The third activity is an end-of-chapter, after-lecture quiz designed to help students retain important facts at their fingertips. Having an internalized set of facts allows them to more readily make connections among concepts. The questions for each chapter are available online as a resource for instructors. Within the course, these multiple-choice quizzes are delivered online, and each student is required to re-take each until he or she scores one-hundred percent (Mastery/Precision learning), without penalty for the number of attempts. Thus, every student gets one-hundred percent, which only shifts the mean of the distribution of the class scores and does not affect the grades. This provides both a disincentive to cheat and some breathing room to encourage self-study.

SUGGESTED READING

Since this book is designed to be focused, rather than exhaustive, I recommend the following books for further reading on big-picture issues, even though some of them are a little old.

1. *A Briefer History of Time* (Stephen Hawking and Leonard Mlodinow, 2005)
2. *Black Holes & Time Warps* (Kip Thorne, 1995)
3. *Stephen Hawking's Universe: The Cosmos Explained* (David Filkin, 1997)
4. *The First Three Minutes* (Steven Weinberg, 1993)
5. *Theory of Everything: The Origin and Fate of the Universe* (Stephen Hawking, 2003)
6. *The Charm of Strange Quarks: Mysteries and Revolutions of Particle Physics* (R. Michael Barnett, Henry Muehry, and Helen R. Quinn, 2000)

For more exhaustive readings, but still at an introductory level, I suggest the following:

1. *The Essential Cosmic Perspective* (Jeffrey Bennett, Megan Donahue, Nicholas Schneider, and Mark Voit, 2010)
2. *Foundations of Astronomy* (Michael Seeds and Dana Backman, 2010)

For more detail see:
An Introduction to Modern Astrophysics (Bradley Carroll and Dale Ostlie, 2007)

Introduction: Big and Small Stuff

Many of us have heard that the universe began with a **big bang**. Inquiring minds want to know what that means, and the evidence for such a claim. As we will see, the evidence points to the universe being much smaller and hotter than it is today. The evidence also indicates that our universe has been expanding rapidly over the last fourteen billion years or so, and that it continues to expand today.

While we do not know for sure how it started, it appears to have burst into existence (perhaps with a big bang?). As we will learn, however, the words "big bang" are a deceivingly simple way to explain an enormously complex process. The story of how our universe became the way it is today is far more beautiful and remarkable than a mere clear-cut description of how it started.

This book will be your guide to our universe and its approximately fourteen-billion-year history. We will describe how it developed from a tiny space into an enormous place with black holes, dark matter, dark energy, and human life. This is one heck of a story, and it will be told without math so that it can be easily understood. More importantly, we will use everyday language to describe some of the evidence that gives scientists confidence in the validity of this story.

To make our topic as fun and interesting as possible—and since we have a long trip together—I have split our story into six parts, which I will call units. They are as follows:

1. Introduction: Big and Small Stuff
2. Physics We Need: General Relativity, Dark Matter, and Quantum Mechanics

3. The Evidence for the Big Bang
4. The Evolution of the Universe: What Happened After the Bang
5. Massive Things: Galaxies, Stars, and Black Holes
6. Early Times, Dark Energy, and the Fate of the Universe

Our first unit consists of four topics. They are:

Chapter 1: Introduction
Chapter 2: Going Big
Chapter 3: Going Small
Chapter 4: Evidence and the Scientific Method

In this unit, we will describe not only what we are trying to explain and understand, but also the thought path and reasoning we will take to get there. The first chapter describes some of the big-picture ideas we want to get across as you read the book. Then, in Chapters 2 and 3, we will describe some of the biggest and smallest things in the universe. We will also look at some of the hottest topics in science today, such as dark matter, dark energy, and black holes. We will come back to these in more detail in later chapters.

Chapter 4 will introduce the use of evidence and the scientific method—the process scientists rely upon to place confidence in our story. In some sense, we are trying to answer the question: What is the story and what is the evidence that led scientists to have real confidence that this is a good explanation of what happened? Ultimately, we are going to tell the history of the universe. But like any good story, we need to start at the beginning.

Introduction

CHAPTER 1

This book is designed to teach you about the mysteries and wonders of our universe. We live in a special time—a time when topics like the big bang and black holes regularly appear in books, blogs and newspapers, and on television programs. These are weighty subjects and we would like to know why scientists have the confidence they do, and how they came to have such confidence.

In many ways, science is the attempt to answer important questions using evidence. Scientists trying to understand the mysteries of the universe are like detectives working to bring a criminal to justice. I like thinking of the famous detective Sherlock Holmes. He works hard to systematically figure out who committed the crime, when it happened, and how the culprit did it. Holmes gathers evidence so that the police, as well as a judge and jury, will be convinced they have the correct perpetrator. Said differently, Holmes works to put together a compelling case.

Another good example is from the not-quite-as-famous television show *Law & Order*. Just in case you're not an avid fan, District Attorney Jack McCoy is an honest lawyer who works hard to understand what really happened in the course of a crime. During a trial, he faithfully uses evidence to lay out the story of how the crime occurred. His objective is to prove to the jury the truth of his words.

While the real criminal justice system is more complicated, we can imagine ourselves as members of the jury, deciding if the lawyer's story is right and whether the evidence presented is persuasive.

One of the most important cases we will consider—and the one that will occupy the bulk of our time—is how the universe came to be the way that it is.

Most of us know a little about the case already. For example, while you have heard that there was a big bang about fourteen billion years ago, you've likely not gathered much beyond that. To be convinced of scientists' claims, there are many questions that will need answers. For example, how did we get from the earliest moments after the bang to the universe we live in today? What evidence supports what scientists call the big bang theory?

Other important mysteries need to be solved as well, and we will talk about many of them. For example, what are black holes? Are they real? Is there any evidence for them? Should I be worried? Is there a reason both black holes and the big bang are in the title of the book, even though they seem unrelated?

There are good answers to these questions—answers we can understand without getting a degree in physics. For a sneak peek at one such answer, I'll go ahead and tell you sixteen chapters early why "black holes" and "big bang" are both in the title of this book. There is good reason to think that the creation of a black hole is like a mini-big bang, but backward in time. For now, think of an explosion in a movie, but with the film running in reverse and all the pieces falling toward the center.

Through creative detective work, we now have a good degree of confidence in many explanations of these mysteries. In numerous cases, there is ample evidence to confidently distinguish among competing versions of the story. Personally, I think everyone should know not only the answers to these questions, but also the supporting evidence. I hope that one day, we will talk about the big bang and black holes the same way we talk about the Earth being round, or the Earth going around the Sun.

While it is clear that scientists know a fair amount, it is equally clear that plenty of mysteries still stump us. For example, what are dark matter and dark energy? We will come back to these in Chapters 6, 15, 18, and 19.

While scientists are not quite ready to bring these cases to trial, the enormous amount of recent progress allows us a glimpse into the formulation of a cohesive case and a conjecture about opening statements to a jury.

It is worth sharing one more idea before we get started. One of the great physicists of the twentieth century, Nobel Laureate Richard Feynman, was fond of understanding things at an intuitive level. While he was one of the most proficient people of his day in advanced mathematics, he is quoted as saying that if you can't explain the essential ideas behind a concept without the mathematics, then you don't really understand it. It is with this idea in mind that we march forward toward a non-mathematical understanding of the universe.

1.1 WHERE ARE WE GOING?

In order to answer big questions about our universe, it is useful to know more about the "stuff" it's made of. This includes things like protons, atoms, stars, and galaxies. It is also helpful to learn a little about how scientists used detective work to discover and understand these things. In some sense, the big things and the little things are pieces of evidence at the scene of the crime that we are trying to explain.

It is also crucial to learn more about two of the great ideas and theories of physics—**general relativity** and **quantum mechanics**—which describe how these various components of our universe are held together. We will cover them more in Chapters 6 through 8.

As we will see, the biggest things in the universe—like planets, stars, and galaxies—are well-described by general relativity, while the smallest—like atoms, protons, and electrons—are better described by quantum mechanics. To understand the universe and the stuff in it, we need both theories. To make it more fun (or complicated, depending on your perspective), it turns out that these theories do not completely agree with each other.

There is a lot here, but don't worry. You are not going to learn all of physics in this small book, and you will not be asked to calculate anything. What we want is for you to understand and be able to re-tell the story to any non-scientist. With that in mind, we start by describing different things in the universe, from the size of a pigeon's nose to the biggest thing we know: the universe itself.

Going Big

CHAPTER 2

Let us say our universe started with a bang—a really big bang—about fourteen billion years ago. If this is true, then fourteen billion years after the bang, the universe would have grown from essentially zero size into the huge universe we see today.

Unfortunately, we've lacked the benefit of a good surveillance camera since the beginning of time to help prove this. So, like a good detective who wants to know what happened during the "crime," we look for clues at the scene. Also like a good detective, we must ask questions such as: What clues are available to us today? What are the pieces of the puzzle we are trying to put together? Does the evidence actually point to a big bang about fourteen billion years ago? Are there other possible explanations that fit all the clues?

We begin our investigation by looking at some things of different sizes in today's universe. Specifically, we start by looking at things of familiar size, then zoom out to sizes that are ten times bigger. We'll continue this "ten times bigger" process a second time, third time, fourth time, and so on.

You can think of it as looking at crime scene photographs from many different vantage points. Each photograph, in this case, is taken at a much further distance than the one before. After we have gone from the sizes we are used to looking at with our eyes to the biggest thing we know—the visible universe itself—we will move to the smallest things in Chapter 3.

2.1 STARTING WITH THE FAMILIAR

To give you a sense of the different sizes out there and just how big the universe is, we will look at a series of pictures, each of which contains a view of a different-sized object. Figure 2.1 shows a series of pictures starting with sizes and things we all know well. Here, we start by looking at something that is about 10 centimeters (about 3 inches) across and 10 centimeters high. We can write 10 centimeters as 0.1 meters, or 0.1 m for short.

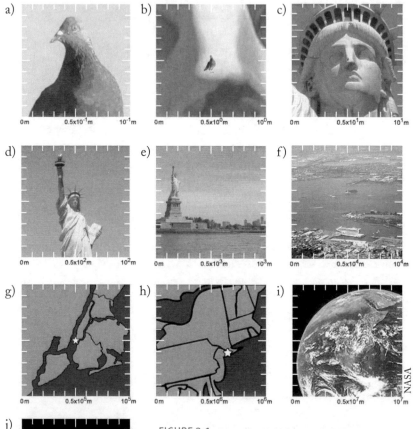

FIGURE 2.1 Pictures with things of different sizes. We start out with sizes we know and love, and then get bigger. Each picture has a set of labels along the bottom and side that indicate the sizes. For example, in the top left we see a size of about 10 centimeters across (10^{-1} m), which is about 3 inches. As we move to figures to the right and down, we get bigger and bigger sizes. By the time we get to the last figure in the bottom left, we are seeing a size that is 10^8 m across, which is greater than the distance across the Earth. Note that the lines in "g" and "h" have been added to the pictures to help us get a sense of the size of the eastern part of the United States.

BIG BANG, BLACK HOLES, NO MATH

In *The Extravagant Universe*, Robert P. Kirshner notes that "a scientist's job is to take something beautiful and turn it into a graph." Note that in these figures (and we will see many more in later chapters), we have done just that to help us visualize information and more effectively tell the story. Each of our graphs typically has both a horizontal and a vertical axis, often to demonstrate the distance between two points. From one side of Figure 2.1a to the other, for example, is about 10 cm across.

A quick aside before we continue: Since we will be talking about very big and very small numbers, we will use scientific notation for convenience. In this case, 0.1 m is written as 10^{-1} m. Box 2.1 has more detail on measurement conversions. It also explains scientific notation, in case you are not familiar with this way of talking about numbers.

BOX 2.1

All the sizes we will talk about in this book will be in the metric system.

1 meter = 100 centimeters ≈ 3 feet

1 kilometer = a thousand meters = 1,000 meters ≈ 0.62 miles

1 megameter = a million meters = 1,000,000 meters ≈ 621 miles

1 gigameter = a billion meters = 1,000,000,000 meters ≈ 621,371 miles

In science, we deal with very big and very small numbers. To handle both extremes, we use scientific notation with powers of 10 to say how many decimal places there are before or after the first digit in our number. For example, $1.5 \times 10^3 = 1,500$ and $3.47 \times 10^{-5} = 0.0000347$. Also, we write meters with an **m**, centimeters with a **cm**, and kilometers with **km**. Same thing for feet **ft** and miles **mi**. So, re-writing the above we have:

1 m = 10^2 cm ≈ 3 ft

1 km = 10^3 m ≈ 6.2 x 10^{-1} mi

Now back to Figure 2.1. As we zoom out to views of 1 m across (about 3 feet) and 10 m across (30 feet), we are still at sizes we recognize. We can now start to see the face of the Statue of Liberty. The next jump is to a view that is 100 m across (about 300 feet). Most of what we can see is the statue, with little of the surrounding area. If we did not know there was a city nearby, we probably could not have figured it out from these three vantage points.

The second row of pictures shows various views of New York City's harbor and the statue's location within it. At a view of 1,000 m (10^3 m or 1 kilometer, also written as 1 km) across, not only can we see the statue, but also we can start to see the city in the background. From this view, we cannot figure out too much about the city or how it was constructed. But, from the height of an airplane in flight, we can actually learn a lot more.

Our view at 10 km (10^4 m) shows some of the streets of Manhattan, and we can make out the grid structure of the city. Being able to see this structure is important because we can learn a lot about the city—how it is organized and how people live and get around. It can also tell us a fair amount about how the city was built many years ago.

Much like looking at Manhattan, if we view the world from the right distances, we can see its underlying structure. And if we understand this structure, we can use it to learn something very important about the physical world—in terms of both big and small things out there—and how it all came to be.

The next picture shows all of New York City, followed by a drawing of New York State. From this vantage point, the grid structure seen earlier would be barely visible. We might not be able to tell there was a city or even human life down there (although looking at night would change our opinion because of the lights). It is not that the grid structure of the city does not exist or that it is not important. It is just that without a detailed picture, we cannot see it. Said differently, the grid structure is important at some sizes, but not at others.

It turns out that this is an important lesson. Just by looking at the structure on one scale, we cannot tell if what we are seeing is an important clue or just an interesting tidbit of information. In order to understand the whole picture, and its individual pieces, we have to see it at many different sizes.

In the last two pictures of Figure 2.1, starting with 10^7 m across, we can see the Earth as the astronauts aboard Apollo 17 did. Based on the earlier scale-picture of New York, the Earth could have been flat, but now from this view, we see something very different—we can now clearly see that the Earth is round.

2.2 GETTING FARTHER OUT: OUR SOLAR SYSTEM

To look at sizes much bigger than the Earth, we've included drawings of what things look like over time. From this view, we can start to see the next order of structure: orbits.

In Figure 2.2, beginning with a view of 10^9 m (a billion meters or about a million miles), we see a drawing with the Earth at the center, and a dashed line that shows the path of the Moon (which is more than eighty times less massive than the Earth) as it goes around the Earth (in 27.3 days). At 10^{10} m, we can see the Earth on its path around the Sun (which is more than three hundred thousand times the mass of the Earth). Also shown is the path of the Moon going around the Earth as the Earth is moving around the Sun.

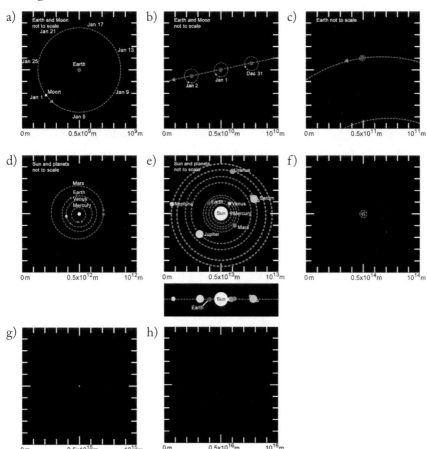

FIGURE 2.2 A view of our Solar System where various things (like the Moon) orbit other things (like the Earth). In the top row, the dot at the center is the Earth, and the dot moving around it is the Moon; the dashed line shows the path of the Moon. As we move out, we can see the path of Venus in the bottom part of the figure. In the middle row, the Sun is now in the center and things orbit around it. We see a view of the paths of the inner planets (Mars, Earth, Venus, and Mercury—ignore the sizes of the planets and Sun in these pictures), and then, out farther, the outer planets. On the bottom row, we see all the planets in our Solar System as they become a smaller and smaller part of our field of view.

By now, we start to notice a pattern of one thing (or "body," as astronomers call them) orbiting another thing. We will notice this again at larger sizes, so we should keep this in mind. When we move to 10^{11} m, we not only see the orbit of the Earth around the Sun, but also we can now see the orbit of Venus. And when we jump to 10^{12} m, we can see the orbits of all of the inner planets.

An interesting observation is that Mars, Earth, Venus, and Mercury all orbit the Sun in the same direction (counterclockwise in this picture). Not only that, but also they seem to move in roughly circular paths. At 10^{13} m, we see the outer planets: Jupiter, Saturn, Uranus, Neptune, and the dwarf planet Pluto (which is so far out that it takes 248 years to orbit the Sun a single time!). Again, all the planets orbit the Sun in the same direction and most of the mass in the Solar System is in the Sun. Are these things a coincidence?

In the same picture, we look at our Solar System from the side rather than from the top. All the planets nearly line up in some sense. The effect is reminiscent of Saturn with its rings. Is this also coincidence? Did this happen by chance, or is there a reason for it? A good detective would suspect that this is a clue to something important. Similarly, we should be intrigued and tantalized just as astronomers were many, many years ago. Perhaps we are on to something.

In the next set of images in the figure, starting at 10^{14} m, the whole Solar System gets to be a smaller and smaller part of our view. At this distance, only a tiny fraction of the space out there is made of large objects like planets and stars. With a view of 10^{15} m, the orbit of Pluto shrinks to a miniscule circle in the middle of our view, and we see more interstellar space (a word astronomers use to mean "the space in-between stars").

By the time we reach 10^{16} m, our Solar System appears as a tiny dot in the middle of what looks like virtually empty space. It is not, however, completely empty. We will come back to describe some of the things that fill it in Chapters 6, 7, 12, and 15 when we discuss atoms, the cosmic background radiation, and a mysterious substance known as dark matter.

2.3 STARS, OUR MILKY WAY GALAXY, AND THE REST OF THE UNIVERSE

Stars (other than our Sun) show up as we move farther out as shown in Figure 2.3. At a size of 10^{17} m (about thirty billion times the distance across the United States, or some five hundred thousand times the distance between the Earth and the Sun), we see some of our neighboring stars: Alpha Centauri A, Alpha

Centauri B, and Proxima Centauri. The latter star is closest at 4.0 x 10¹⁶ m from the Sun. To get a sense of the relative distances using a baseball analogy, if the Sun were at home plate on a baseball diamond in New York, and the Earth were 90 feet away at first base, these stars would be located roughly 5,000 miles away in the African city of Timbuktu.

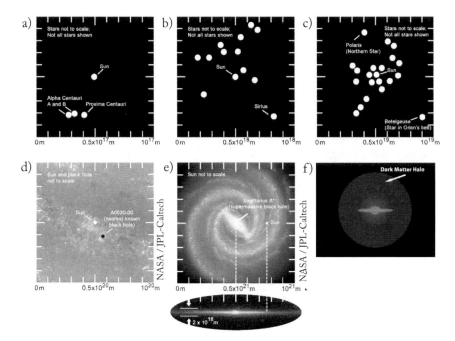

FIGURE 2.3 While no spacecraft has gone far beyond our Solar System, we can infer sizes from the stars nearest our Sun to the full size of the Milky Way. The first row shows the location of some of the brightest and most famous stars in the Milky Way (ignore the size of the dots). The location of the closest black hole known to us is indicated in "d." The top-down view and a side view of our galaxy, in "e," (color version on page C-1) has many of the same features as our Solar System. Note that we are not showing the dark matter, only the stars. The location of the dark matter is shown in "f."

At 10¹⁸ m, there are many more stars. The picture only shows the brightest ones to make things easier to draw; there are actually about two thousand in total. Zooming out to 10¹⁹ m, you'd find about two million stars. We only show a few of these stars and point out two of the most well known.

By the time we reach 10^{20} m, we begin to see our galaxy: the Milky Way. From this view, there are so many stars—about a hundred billion in total—that they all seem to blur together. At 10^{21} m, we see the entire Milky Way. Our galaxy consists of a huge number of stars near the center, known as a "bulge" of stars, and a flat disk outside it containing long "spiral" arms. These arms, in turn, contain younger and brighter stars. Our Sun is located near the center of one of the spiral arms, about one-third of the way inward from the edge. Each star orbits the center once every several hundred million years.

Let us pause to observe our galaxy. One of the first things we notice is that it is similar to our Solar System in a number of ways. For example, the top-down view shows a bulge in the center (like our Sun) with objects orbiting it (like our planets). The side view illustrates that things in orbit are mostly in a flat plane (again, like our planets). We also note that most of the mass in a galaxy is in dark matter which essentially surrounds the galaxy like a giant ball, but emits no light so we can't see it directly. We will talk about some of the evidence for dark matter in Chapter 6.

To be sure, there are many things that are different, but the similarities are re-markable. Is this also an important clue? Why do we keep seeing this same pattern? Is there something that Saturn, our Solar System, and our galaxy have in common? The plot has thickened and we will come back to these questions in Chapter 15. For now, we will give away a little bit of the answer: our Solar System and our galaxy look similar because gravity holds them together in similar ways.

It was only in the 1920s that scientists clearly understood that some of the dots of light they could see in the sky were not distant stars, but were separate galaxies. Before that, most people thought the Milky Way *was* the entire universe. But now, with ever more powerful telescopes and satellites, we can see what they could not. With a view of 10^{22} m, Figure 2.4, we can see that not only are there more than two galaxies, but also there are many galaxies. Interestingly, there are about as many galaxies in the visible universe (10^{11}) as there are stars in our Milky Way, and some of them are extremely far away.

At 10^{23} m, we see that many of the galaxies are often near each other. Astronomers call these "groups" or "clusters." Groups consist of just a few galaxies, while clusters can contain thousands of them. Our "Local Group"—the collection of galaxies of which the Milky Way is a part—is on the fringe of a set of galaxy clusters called the "Local Supercluster."

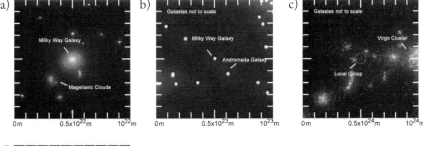

a)

Milky Way Galaxy

Magellanic Clouds

0 m   0.5x10²²m   10²²m

b)

Galaxies not to scale

Milky Way Galaxy

Andromeda Galaxy

0 m   0.5x10²³m   10²³m

c)

Galaxies not to scale

Virgo Cluster

Local Group

0 m   0.5x10²⁴m   10²⁴m

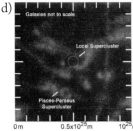

d)

Galaxies not to scale

Local Supercluster

Pisces-Perseus
Supercluster

0 m   0.5x10²⁵m   10²⁵m

FIGURE 2.4 Moving farther out, we see an image of our galaxy surrounded by other nearby galaxies. We stop at 10^{25} m across because we do not know the true size and shape of the observable universe.

When we get to 10^{24} m and 10^{25} m across, we can see the full complement of galaxies that comprise our visible universe, and we note that the most distant galaxies are moving away from us very quickly (more about this in Chapters 10, 13, and 18). At this point, our ability to measure is not good enough for us to discern the true shape and size of the universe. We cannot go out any farther than this, and in later chapters, we will see why.

Now that we have gone from the size of the nose on our pigeon's face to the size of the observable universe, Chapter 3 will take us the other way—all the way down to the very smallest things that have been "observed" in our universe.

Going Small

CHAPTER 3

The time has come to look at the small things. While objects of small size are important in their own right, we will see that large numbers of small things, like atoms, put together in special ways are what make up the stars and galaxies, as well as the human body. Again, we start at sizes we know and then get smaller and smaller until we are limited by technology in our ability to go any further.

3.1 GOING FROM SIZES WE KNOW DOWN TO THE SIZE OF ATOMS AND QUARKS

We start with a simple object: a baseball (Figure 3.1). Moving in to 1 cm (10^{-2} m), we see a single dime. Zooming down to 10^{-3} m (1 mm) we start to see things we cannot easily see with the naked eye: in this case, a dust mite and, at 10^{-4} m, a side view of a single human hair.

At 10^{-5} m, we have a close-up view of a single red blood cell, and we have entered the realm of biology and the microscopic. When we get to 10^{-6} m, we see some human immunodeficiency viruses (HIV), which are known to be one of the larger viruses; and at 10^{-7} m, we see polio viruses, known to be one of the smaller ones. When we reach 10^{-8} m, the double helix structure of DNA becomes visible, and we have entered the realm of chemistry.

At sizes of 10^{-9} m (1 nanometer or 1 nm), we enter the realm of **atomic physics** and notice that an atom is composed of electrons and a nucleus. Despite what we have drawn, our eyes cannot see objects this size in the conventional sense. For example, while the picture makes an atom resemble a balloon, we note that what looks like the surface of the balloon represents the places the electron can go as it "orbits" the nucleus at high speed.

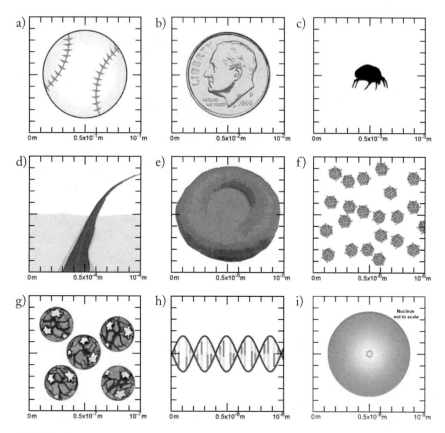

FIGURE 3.1 What things look like at smaller sizes as we zoom in for closer views. In order, these objects are: a baseball, a dime, a dust mite, a human hair, a red blood cell, human immunodeficiency viruses (HIV), polio viruses, a DNA strand, and an atom. It is important to point out that while the picture of the atom in the bottom right looks like a balloon, the tinted surface of the atom only indicates places its electrons are allowed to go. Also, note that the nucleus at the center is not drawn to scale.

Think of how the blades of a ceiling fan or a hubcap on a tire give the appearance of a solid object when they rotate quickly. In some ways, an electron going around a nucleus (the proton is about two thousand times more massive than an electron) is like a planet orbiting the Sun. But instead of just moving in a single circle in a plane, the region highlights the area in which the electron is most likely to be found.

How electrons move around the atom (and it is more complicated than quick movement alone) and why this is important will be discussed in Chapter 7 when we talk about **quantum mechanics**. To move further inside this atom, let's look at a set of snapshots in time.

At a size of 10^{-10} m, or about ten billion times smaller than the height of a person, we come to see the difference between things that are **fundamental** and those that are **composite** (see Figure 3.2). Things that are fundamental are so small that despite years of trying, no one has successfully pried one apart to see if there is anything inside, nor has anyone been able to measure their size. Electrons are known as fun-

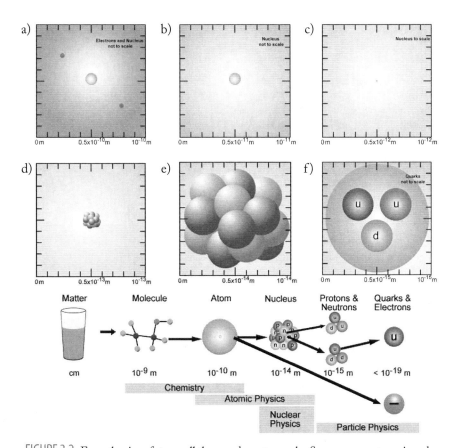

FIGURE 3.2 From the size of atoms all the way down to quarks. Some comments are in order regarding these pictures. First, each illustration shows a simplified "snapshot in time." While the nucleus resides at the center of an atom, the electrons do not ever reside in a single place, as we will discover in Chapter 7. It is equally important to not be misled by the size of the dots in the top row; the dots representing the nucleus and electrons are not drawn to scale. For now, one can think of each dot as an example of the distance each electron can have from the center of the atom. The size of the nucleus, protons, and neutrons are correct in the second row. The rightmost picture in the second row shows three quarks within a proton; again, they do not reside in a single place in space and their size in the picture does not represent their actual size. All we know for sure is that electrons and quarks, if they have any size at all, are smaller than 10^{-19} m across. Color version of the bottom plot on C-2.

damental, and we refer to all fundamental things as **particles**. An atom is described as being composite because it is made up of other things: a nucleus and electrons.

In this snapshot, the ball in the center of the atom represents the nucleus' location within the atom (but not its size), and the dots around it represent the approximate positions of the electrons (again, not their actual size). When we zoom in to 10^{-12} m, the size of the dot represents the size of the nucleus. The electrons are now completely outside our field of view. We are in the realm of **nuclear physics**.

When we reach 10^{-13} m, we clearly see that the nucleus, like the atom, is not a fundamental object, but rather a composite object made up of what we call **protons** and **neutrons,** and that they are held tightly together. At 10^{-14} m, we show a carbon nucleus and we can clearly see the six protons and six neutrons.

Are neutrons and protons fundamental or composite? Before the 1960s, they were assumed to be fundamental and had always been referred to as particles. We now know they are actually composite, although we still call them particles.

Moving down to sizes of 10^{-15} m, we see inside one of the protons and discover that it is composed of three separate fundamental particles, each of which is known as a **quark**. The same is true of neutrons. In fact, both are made up of two different types of quarks, known as **up quarks** and **down quarks**.

Since we are down to the fundamental particles, and in the realm of **particle physics**, we stop here.

To give you an idea of size differences, if an atom were the size of a large city like Los Angeles, then neutrons and protons would be the size of a person, and electrons and quarks would be smaller than a tiny freckle on a child's face.

There have been many attempts to determine the exact size of quarks and electrons, or decide if they have a measurable size at all. The best experiments only tell us that if either has a size, it is smaller than 10^{-19} m.

3.2 OTHER SMALL THINGS

Scientists have great confidence that much of the "stuff" in the universe that we talked about in Chapter 2, like the Earth and stars, is built of just three types of fundamental particles: up quarks, down quarks, and electrons. Essentially, they are what we call the **fundamental building blocks of nature**.

On the other hand, scientists do not yet know what the dark matter we mentioned in Chapter 2 actually is. It is possible that it is a fundamental particle, but there is not yet enough evidence to tell. What we do have confidence in is that it is not made up of known particles.

In addition to the fundamental building blocks of nature, it turns out that there are many different kinds of particles that we do know about and call fundamental. Table 3.1 gives a full list. While we don't know if they are bigger or smaller in size than an electron, we do know that some are lighter and some are much heavier.

Quarks	Leptons	Force Particles
Down	Electron	Photon
Up	Neutrino (electron type)	W
Strange	Muon	Z
Charm	Neutrino (muon type)	Gluon
Bottom	Tau	
Top	Neutrino (tau type)	

TABLE 3.1 A list of the known particles that are believed to be "fundamental." They are grouped into three types: quarks, leptons, and force particles. We will come back to these types in later chapters. We note that there is now significant evidence that a Higgs boson has been discovered. For more information about it, see the Suggested Readings.

As detectives, we note that there are six different types of quarks, although it's not clear why so many. Also, it turns out that there are five other particles similar to the electron (again, for a total of six); we call these **leptons**. We do not really know why we need six of each, or if there are more to be found. But we do know that the fact that there *are* six plays an important role in the earliest moments of the universe.

Another important piece of the fundamental particles story is the existence of **anti-particles**. We call protons and electrons **matter**, and the anti-particle versions of these are called **anti-matter**. Despite being a regular part of science fiction TV and movies like *Star Trek*,[1] these particles are very real. Every type of particle that can exist in nature, like the ones in Table 3.1, has an anti-particle version. For example, there are anti-electrons, often called **positrons**, as well as anti-quarks.

Anti-quarks can be combined to form an anti-proton, and when an anti-electron orbits an anti-proton, we have an anti-atom. Why the universe consists of enormous

---

[1] This includes both engines and anti-matter weapons.

numbers of electrons and precious few positrons (and the same for protons and anti-protons) is an important mystery scientists would like to solve.

One of the important properties of particles, both fundamental and composite, is that they can be **stable** or they can **decay**. When we say that a particle is stable, what we mean is that it can live forever unless something happens to it. For example, an electron exists without end unless it hits something, like an anti-electron. On the other hand, a muon, a particle shown in Table 3.1, will decay after a tiny fraction of a second.[2] Similarly, atoms like hydrogen are stable, but some atoms, like uranium, can decay into helium and thorium.

Before concluding, we note that although scientists have made great headway in learning how the world of the small is structured, there are many questions that remain unanswered. For example, we are trying to figure out whether electrons and quarks are truly fundamental. Also, we want to know whether dark matter is a fundamental particle or if there are other new particles just waiting to be discovered. We will return to these questions in Chapter 19.

We have now described the biggest and smallest things that we can measure in the universe, but we wish to explain how they came to be that way. To do so, we need to look at the scientific method and some early attempts to explain our universe.

In the next chapter, we will take on some of the "facts" that we were taught as young children in school, such as "the Earth is round," "the Earth goes around the Sun," and "our Sun is not the center of the universe." (Don't worry—wse still think these things are true!) But, as good detectives, we need to ask what the evidence is for these conclusions.

Thousands of years ago, the average person thought the Earth was flat, so what do we know that they did not? That one is easy to answer because we can see it is round in Figure 2.1. Others are harder, like: Why do we think we are not at the center of the universe? How did we come to learn this and what is the evidence for it?

---

[2] Remember that the way that particles decay is not the same as they way an old piece of fruit decays. You can think of a particle decaying as the particle simply turning into a set of different particles. This is one of the important predictions of quantum mechanics, which we will discuss in Chapters 7 and 8.

BIG BANG, BLACK HOLES, NO MATH

Evidence and the Scientific Method

CHAPTER 4

As scientists and detectives, our search for answers to questions about the world often begins by looking for clues and using our intuition to piece the puzzle together. While we may have a lot of personal experience with the world around us, unfortunately, our experience with the things we can see or touch can be a lousy or incompetent guide as we try to understand the bigger and smaller things in the universe.

Chapters 2 and 3 emphasized that nature is incredibly more complex and different than what our eyes tell us. For example, we still talk about the Sun "rising" in the east, even though we know the Earth is both rotating around its axis and orbiting the Sun. Unfortunately, history shows that what we *believe* (or our preconceived notions) can prevent us from understanding how things *really* work. There have been numerous "explanations" of nature which have impeded progress toward understanding the universe and how it works.

In this chapter, we will describe the scientific method with an abridged version of how some of the great detectives (shown in Figure 4.1) explained the motion of the planets and eventually came to understand that the Earth is not the center of the universe. Rather than focus on the history, we will focus on how they used evidence and the scientific method to help determine a more correct description of nature. As we look toward later chapters, we will need to use evidence to both build up correct theories of what is going on and also to reject false theories.

Before we begin, it is worth mentioning that while we often talk about the "scientific method," there is no single way to do detective work. Sherlock Holmes certainly could not tell you an exact process for solving crimes. Instead, he took guesses and checked to see if the clues fit, and he used trial and error to gather new facts to build up a case.

Since every case is unique, the methods required to solve each one will present different details. Science writer Isaac Asimov is quoted as saying, "The most exciting

| Claudius Ptolemaeus (Ptolemy) | Nicolaus Copernicus | Galileo Galilei | Isaac Newton |

© RMN-Grand Palais / Art Resource, NY

Image © Oleg Golovnev, 2013. Used under license from Shutterstock, Inc.

Image © Georgios Kollidas, 2013. Used under license from Shutterstock, Inc.

Image © Nicku, 2013. Used under license from Shutterstock, Inc.

FIGURE 4.1 Some of the great early scientists. Each made crucial breakthroughs in the field of cosmology.

phrase to hear in science, the one that heralds new discoveries, is not 'Eureka!' (I found it!) but 'That's funny ...'" Ultimately, the tried-and-true method of logically and systematically testing ideas with data has proven to be extraordinarily productive.

4.1 THE EARLY STUDY OF COSMOLOGY

Cosmology is, in simple terms, the study of the origin and nature of the universe and how it changes over time. People have been trying to understand cosmology in one form or another for ages.

Thousands of years ago, early astronomers looked up at the sky and noticed that almost all of the white dots of light—what we now call stars—moved together over the course of the night. They saw patterns in the stars and called these constellations (shown in Figure 4.2).

They also noticed that five of these lights moved in a strange manner over the course of weeks and months. They called these "planets," which evolved from the Greek word for "wanderers." We now know them by their Roman names: Mercury, Venus, Mars, Jupiter, and Saturn. What made them strange is that they wandered slowly *through* the constellations of the zodiac.[1] It was as if the planets were visiting the stars, one after the other.

What made the planets' motion even stranger was that they looked like they would periodically stop, go backward through the constellation for awhile, and

---

[1] There are only twelve constellations in the zodiac, which are well known to people because of their use in astrology. They are, in order, Aries, Taurus, Gemini, Cancer, Leo, Virgo, Libra, Scorpio, Sagittarius, Capricorn, Aquarius, and Pisces. For more on the difference between cosmology and cosmetology, and between astronomy and astrology, see Box 4.1.

24

BIG BANG, BLACK HOLES, NO MATH

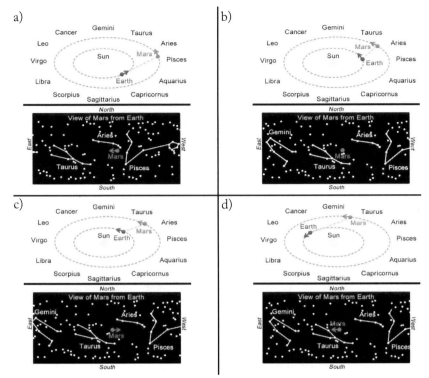

FIGURE 4.2 Two ways of looking at Mars, and the stars (and constellations) behind it, from the Earth. The dotted lines represent the path that the planets take. From the Earth's perspective, Mars looks like it moves backwards for a while before moving forward again. This is known as retrograde motion.

BOX 4.1

This book is about **cosmology**, which is often defined as the study of the origin and nature of the universe and how it changes over time. Unfortunately, the word cosmology is sometimes confused with the word **cosmetology**, which is the study of cosmetics and their use. I wish I could say that this did not happen very often, but it does. Another easily confused pair of words is **astronomy** and **astrology**. **Astronomy** is the study of stars and other objects in the universe. **Astrology**, however, has nothing to do with scientific study. It is the belief that the position of the stars and planets influence human affairs and events. For example, you can look at Figure 4.2 and see what it means for "Mars to be in Aries." Many people find looking at their horoscope in the newspaper more fun than looking at the stars, perhaps because they do not understand what stars really are, or how astrologers make their predictions.

then start up again along their original paths. This process of stopping, moving backward, and then proceeding forward is known as **retrograde motion**.

In the second century, the astronomer Claudius Ptolemaeus—or "Ptolemy," as he is more often called—pulled together the best evidence of the day and laid out his description of the universe. In it, the Earth was the center of motion and the Sun, Moon, stars, and "wanderers" orbited it. The explanation of the retrograde motion was that as the planets traveled around the Earth in a circle, their paths would include a small, extra circle around the main path.

While not a perfect analogy, think of sitting on the Sun and watching the Moon as it goes around the Earth and the Earth as it goes around the Sun. The path of the Earth is the main circle and the orbit of the Moon around the Earth is called an **epicycle.** Next, pretend that you can't see the Earth; the Moon going around the Sun by itself in a funny, almost-orbit-like motion. Ptolemy's description had very large epicycles and provided a reason why we observed the retrograde motion of the planets. However, there was neither a convincing explanation of why the epicycles existed—the planets weren't orbiting anything like the Earth in our example—nor was there an evidence-based connection between this motion and the motion of the stars.

Although the details changed over time, in Europe, this basic description was generally accepted for about fifteen hundred years. It is not too hard to understand why: The Earth does not feel like it is moving, and from our vantage point on the Earth looking up at the sky, the stars, planets, Sun, and Moon all appear to go around us.

4.2 CRACKS IN EARLY COSMOLOGY

In the early 1500s, Nicolaus Copernicus published a new description of the motion of the planets that contained many elements commonly accepted today.[2] For example, although the Sun and planets may *look* like they are orbiting the Earth, the Sun is actually at the center of the motion and the planets and the Earth orbit around it.

This has the appeal of simplicity; the Earth and the other planets all move in simple, orbiting paths. The data only look complicated because of our vantage point here on the moving Earth.

---

[2] To be fair, many ancient peoples like the Greeks postulated that the Earth might be going around the Sun. However, they ultimately decided against it for a number of reasons. The history here is very interesting and I encourage you to follow up on it online or in the Suggested Reading texts. We also note that Copernicus had some parts wrong. For example, he thought the main orbits were circles with epicycles, although we now know they are really ellipses.

Copernicus' description also provided a straightforward explanation for the apparent retrograde motion of each planet: it is just the Earth passing the planets as they travel on their individual orbits around the Sun (see Figure 4.2).

This brings us to a crucial question: If there are two competing explanations that can each account for all the data, how do we pick one? Both of these explanations, for example, were equally adept at predicting the location of the planets at any moment in time.

There is nothing wrong with picking the explanation you like better—but only if there is no evidence against it. That said, there is real benefit to choosing the explanation that makes the fewest new assumptions or is the simplest. This idea is known as **Occam's razor** (pronounced AHK-uhmz RAY-zuhr). Should the people of Copernicus' time have accepted his description as true? Perhaps not. It is true that the modern view is *simpler* in some ways (retrograde motion is just planets passing each other), but not in others (requires a rotating Earth).

A more logical way to answer the question is to ask about the evidence for and against the two ideas. Consider a question about the modern description. Isn't there evidence against the idea that the Earth is rotating while it orbits the Sun? We don't *feel* like we are rotating.

I can tell if I am on a merry-go-round. Why don't we fall off the Earth like an ant placed on a quickly spinning bicycle wheel? Why don't we always feel wind in our face when we are pointed in the direction of rotation, and at our backs when we face the opposite way? If I jump into the air, why doesn't the Earth rotate under me? We now know the answers to these questions,[3] but the people of Copernicus' time did not.

A better process is to gather new evidence for or against one description or the other. This is eventually what happened—but it required tools that had not yet been invented.

In the early 1600s, many astronomers, including Galileo Galilei, started gathering data using a powerful new tool: the telescope. With a telescope, astronomers could now see things previously invisible to the naked eye. Galileo made two key observations, shown in Figure 4.3, which offered the first "proof" that not everything orbited the Earth.

---

[3] This will be discussed more in Chapter 6 when we cover gravity.

Jupiter: NASA. Io: Galileo Project/JPL/NASA. Europa: NASA/JPL/DLR. Ganymede: NASA. Callisto: Adapted from Image © Antony McAulay, 2013. Used under license from Shutterstock, Inc.

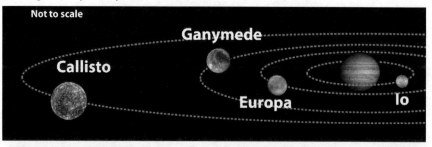

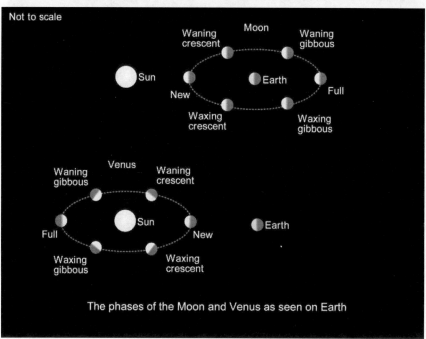

FIGURE 4.3 Two drawings depicting the views from a telescope that changed our understanding of the Earth and our Solar System. On the top, the moons of Jupiter observed by Galileo provided evidence that not all things orbit the Earth. On the bottom, light from the Sun bounces off both Venus and the Moon and toward the Earth. This causes "phases" in the amount of light we see reflected off each. As shown on the bottom, we are able to see all the phases of the Moon because of the relative locations of the Earth, Moon, and Sun. From our vantage point on Earth, we are only able to see all the phases of Venus if Venus goes around the Sun and not around the Earth. It is worth noting that the phases of the Moon are due to reflected sunlight, not the Earth casting a shadow on the Moon—that is a lunar eclipse.

Galileo's first observation was that Jupiter has moons that orbit it. Secondly, he was able to see the way sunlight reflected off Venus and toward us here on Earth as both the Sun and Venus moved through the sky. The observation of the full set of Venus' "phases" (like the phases of the Moon) could only be explained if Venus orbited the Sun—there was no explanation in the complicated epicycle-laden path in the geocentric model.

Over time, astronomers added precise measurements for the locations of the planets, as well as their brightness, which suggested that if the planets went around the Sun, then their orbits were shaped more like ovals (ellipses) than circles.

The matter was not considered settled until the late 1680s. Sir Isaac Newton developed his theory of gravity, which explained *why* the planets move the way they do. He described the motion of a planet around the Sun as the result of gravitational force between the two "heavenly bodies" (more about gravity in Chapter 6).

Gravity stipulates that the same force that pulls an apple from a tree to the ground also pulls the Earth toward the Sun and keeps it in orbit. This explained the paths of the planets around the Sun, why the paths were ellipses, why they moved as fast as they were observed to move at any point in the sky, and why they had a particular brightness. For example, the Earth goes around the Sun faster than does Mars because the Earth is closer to the Sun and the force on it is stronger.

Newton's description of gravity stood unchanged for centuries. But in the early 1900s, a young German scientist named Albert Einstein introduced a more refined and sophisticated version of gravitational theory: general relativity.

4.3 THE SCIENTIFIC METHOD

With this important historical example behind us, it is time to move to the question of how we separate the stories that are true from those that are not. Like detectives, we need evidence and a good **theory**—a good *scientific* theory, to be exact. Since scientists often use the word "theory" differently than it's used in everyday conversation—and since they often use words like **model** and **hypothesis**—it is worth clearing up these differences.

We start by defining a hypothesis as *a proposed explanation for the data*. In some ways, you can think of a hypothesis as a guess. For example, if there is a broken jar on the floor and a guilty-looking cat sitting next to it, a reasonable hypothesis is that the cat did it; "the cat-did-it" hypothesis accounts for all the initial data.

Then again, it might be useful to construct a simple description of what happened. This is what scientists would call a "model." For example, a decent model of what happened is that the cat jumped onto the table and knocked the jar onto the floor, breaking it. This model would account for all the data. However, it suggests that other useful data should also be gathered.

If we notice that the jar was very heavy—in fact, too heavy for the cat to knock off the table—we say that our hypothesis is false. However, if it was not too heavy for the dog to knock over, we look for a way to test a "dog-did-it" model. If we look more closely at the broken jar pieces and find a dirty dog print on one of them and no cat hair or paw prints, our new data supports the "dog-did-it" hypothesis/ model and further contradicts the "cat-did-it" hypothesis/model. Most cats are curious, so it is not unexpected for a cat to show up after the jar fell. Thus, we have a reasonable explanation for the "cat-at-the-scene" data.

This is the essence of how we use the scientific method. For more on other important issues—like repeating an experiment, making predictions, or understanding how science progresses over time, see Box 4.2.

We next move to the word "theory." So many complicated ideas are contained in this one word that I do not have a good, simple definition for it. To help explain what I mean, let's start with how it is used incorrectly in everyday conversation.

If someone said to me that the reason the Chicago Cubs baseball team will never win the World Series is because they are cursed, I might shrug my shoulders and glibly respond, "It's a theory." However, that's not really right. A proper scientific response would instead be something like, "That's a hypothesis, but I'm not sure how to test it." Scientists reserve the word "theory" to describe statements that are much more generally true, like "gravity attracts things to each other." In some sense, it is a much bigger—or all-encompassing—statement.

For a hypothesis to be elevated to the status of a theory, it must not only pass many tests, but also it must make large numbers of predictions that can be tested in future experiments. If it makes no predictions, then it can't be granted the title of "theory." If it does make predictions, and experimental results disagree with the predictions, then the theory is false. Indeed, for anything to be elevated to the status of a theory, it must be falsifiable.[4] For us to have confidence that a theory might correspond to truth, scientists agree that a good scientific theory is one that has made enormous numbers of predictions and none have been falsified by experimental testing. This is a pretty high standard for calling something a theory, and it is the framework on which science is built.

---

[4] To illustrate what we mean by falsifiable, we go back to our broken jar example. The cat-did-it and dog-did-it hypotheses are falsifiable in the sense that one can make predictions about clues that could be found at the scene of the crime, like the size of the paw prints on the broken jar. There is no way to confirm or falsify a ghost-did-it hypothesis.

For a good theory, I look for three main ingredients:

1. It must accurately describe a large class of observations, clues, or other pieces of evidence.
2. It must make definite predictions about the results of future observations or experiments.
3. It must be falsifiable and have passed all experimental tests.

BOX 4.2

If a group of people stand outside on a clear night looking up at the stars, I might tell them, "Take a look at that flickering dot of white light up there in the dark night sky. That is a giant ball of hydrogen that is both incredibly hot, and billions of miles away." I have done this a couple of times with both kids and adults, and most people shrug and say, "Okay." They largely accept the idea without questioning it too much. The fact that I am a scientist makes people more likely to believe me. It is also likely that they have heard someone else tell them the same thing before on TV or in science class. They might even say that they "heard it on TV" as a way of implying (erroneously) that it is "corroborating evidence." Then again, if that same group of people stood outside near a park bench a few feet away and I said, "Don't sit on that bench—it has wet paint," most of them would not take my word for it; they would go closer and maybe even touch the bench to see for themselves. The same is true when a waiter brings me a plate at a restaurant and says, "Don't touch it; it's hot." I am never able to resist touching it just to see for myself.

Going over and touching a star is much harder than touching a park bench or a hot dish. Continuing with the analogy is useful. It is reasonable to ask why I believe the bench has wet paint. What is the evidence? If I reply that I was looking at the bench from 1,000 feet away and I thought I saw drips, you would be skeptical of my ability to see that clearly. Then again, if I said I saw someone sit down on it and, when they got up, they then had paint on their clothes, you might agree that this is good evidence. In the same way, if I tell you about stars from what I see with my naked eye, you should be skeptical. But if I use powerful telescopes, teach you how they work, and tell you the data I saw, then you might start to accept my description as being credible. Ultimately, scientists have built many powerful telescopes and check on each other to make sure their ideas and concepts of how the world works are backed up by lots of different types of evidence. Ideally, you would come join us by looking for yourself.

If we do an experiment and the results disagree with the predictions of a theory (assuming the experiment was done correctly), then the theory is wrong or must be modified. It has been said that even the most beautiful theory can be slain by a single ugly fact.

But if we can make accurate predictions for an experiment that has never been done—and the data agrees with the prediction—we have good reason to believe that there really are underlying principles in our world, and that our theory has *some* truth to it. Science and the scientific method do not promise eternal truths, but only the systematic elimination of false hypotheses. Theories, therefore, are the embodiment of our current best explanation. For more, see Box 4.3.

BOX 4.3

It is important to say a little about what happens to theories as science progresses. Some people believe it is not worth studying physics because it is just a matter of time before all the current theories are shown to be wrong and/or need to be replaced. For example, Ptolemy's theory held sway for fifteen hundred years until Newton; as we will see Newton's theories were replaced by Einstein. Why study Einstein's theory if it is just going to get replaced? Is it reasonable for a great deal of money to be spent on science if the theories might someday get replaced?

It is important to remember that a good scientific theory is not "truth," but just our best understanding of what is going on. It is our best explanation of all the data at the moment. As time goes by, only those theories that explain all the available data survive. Thus, it becomes less and less likely that a theory that has survived very detailed experimental testing might later become invalidated. On the other hand, it is quite likely that the theory may need to be *extended* in the future to simultaneously explain new and more detailed experimental results as they come in. In other words, we don't usually expect to "throw away" descriptive ideas that have already proven to work very well; we may, however, need to add to them so that they can start working even better. Along the way, we may hope to gain a better understanding of *why* the original theory worked so well in the first place. For example, while Einstein's theory may have replaced Newton's description, Newton's description is much easier to use, and is more than good enough to use to build bridges and buildings. This is why large numbers of engineering students still learn it rather than Einstein's version. In some sense it is incomplete, but if you want to predict the outcome of an experiment where things move slowly, then it does a great job!

4.4 When Theories go Bad and when Weird Theories make Good

At some point, instruments and experiments get powerful enough to observe things that our eyes cannot see directly. We can use a microscope, for instance, to view minuscule things and a telescope to view the faint and/or distant. In the same way that Ptolemy's model couldn't explain what Galileo saw in his telescope, what happens if a theory cannot accurately predict what we see in a microscope? In this case, we have to try new—and often weirder—theories.

What if we come up with a theory that accurately predicts the outcome of an experiment? That would not be too bad. But what if we do not *like* the theory or it strikes us as weird? What if it goes against our intuition or what our parents or teachers taught us? Well… maybe we would accept it and maybe we would not.

What if, after studying our weird theory for a while, we realize that it makes further weird predictions about what we would see in an experiment we have not yet done? And what if we do the experiment and actually get the results that the weird theory predicted? At what point do we start believing there is some truth to the theory?

Sometimes we do not want to believe what the data tell us. It is like knowing that your neighbor has been murdered, and then being told by the detective that the evidence is mounting against your close childhood friend. Even if your friend has always been kind to you, it may be that he is the killer. To quote the world's most famous detective, Sherlock Holmes, "When you have eliminated the impossible, whatever remains, *however improbable*, must be the truth."[5]

Scientists tend to have confidence in theories that do a good job predicting the results of experiments. They also test them repeatedly in new and clever ways—just in case. While many of the theories of physics seem strange to most people, they did not get *weird* until the early part of the twentieth century. It was during this time that scientists started coming up with new theories, such as general relativity and quantum mechanics. In the next chapters, we will start to de-mystify these theories. Onward to Unit 2!

---

[5] Sir Arthur Conan Doyle in *The Sign of Four.*

Physics We Need: General Relativity, Dark Matter, and Quantum Mechanics

In order to understand how the universe has changed from its earliest beginnings (what scientists often call the "big bang") until today, we will need to learn some physics. Doing so will also help us understand the evidence that has given scientists confidence in this description.

To explain the need for this process, I once again turn our focus to crime shows. I don't know much about human anatomy, but I know enough to appreciate that human beings have unique fingerprints. I also know that detectives can use this important fact to help solve crimes. To solve the mysteries of our universe, we, too, need to know the kinds of factual tools that will help us to gradually unravel the puzzle of its origins.

The heart of the science we need to grasp is encapsulated in the two great theories of physics: **general relativity** and **quantum mechanics**. Here, we are using the word "theory" in the broadest sense from Chapter 4. We will not devote the years professional scientists spend mastering these theories, but will instead focus on a conceptual understanding you can use here and now. While each theory is subtle, complex, and somewhat mysterious, both are fascinating and beautiful in their own right. Understanding them will illuminate not only the topic at hand, but also other exciting things in our universe like black holes, dark matter, and dark energy.

The ideas of quantum mechanics and general relativity are spread throughout this unit. As a whole, the unit has five chapters, each of which covers a set of separate, but interrelated, topics. They are:

Chapter 5: Light and Doppler Shifts

Chapter 6: Gravity, General Relativity, and Dark Matter

Chapter 7: Atomic Physics and Quantum Mechanics

Chapter 8: Nuclear Physics and Chemistry

Chapter 9: Temperature and Thermal Equilibrium

In many ways, these ideas form the foundation for our current understanding of science, in particular, the physical universe. Because there is no perfect or required order in which to present these subjects, we will start by describing light since we use it to "see" the universe.

Light and Doppler Shifts   CHAPTER

A large amount of what we have learned in science comes from what we see when we "look" at things. The act of looking, for most living species, is what occurs when light coming from something hits our eyes. We know the Sun exists because we see the light coming from it. We see the walls in a room around us because light bounces off them and into our eyes. Similarly, we know about atoms because of the light we see that is emitted from them, or bounces off them.

There is a lot more to light than meets the eye. For example, there are many different kinds: some that we can see with our naked eyes, others that we cannot. Many people know the phrase "at the speed of light" and recognize that it is a really fast speed.

To truly understand light, though, you must first realize that it acts as both a wave and a particle. Since most people do not typically think about light in this way, we will start by talking a bit about what it means to be "wave-like" and "particle-like." After that, we will discuss the different things we can learn from each.

Since light behaves like a wave, it can undergo what scientists call the **Doppler effect**. By understanding and utilizing this phenomenon, we can figure out whether a star or galaxy is moving toward us or away from us. Using properties of both particles and waves have allowed scientists to conduct better experiments and gather more clues about the universe. It is remarkable how much we can learn about the little white dots in the sky by looking at the light that comes from them.

5.1 LIGHT AS A WAVE

Since the light we see with our eyes acts like a wave, it is important to understand what we mean by "wave." Most people have been to the beach and seen waves as they come into the shore (see Figure 5.1). It is not hard to see the peaks (where the water is above the usual height of the water level) and troughs (where it is below). When scientists talk about waves, they do not talk about the individual wave that crashes on the beach, but rather the amount of space in between the corresponding parts of the waves; for example, between the peaks of two neighboring waves. This distance is known as the **wavelength**.

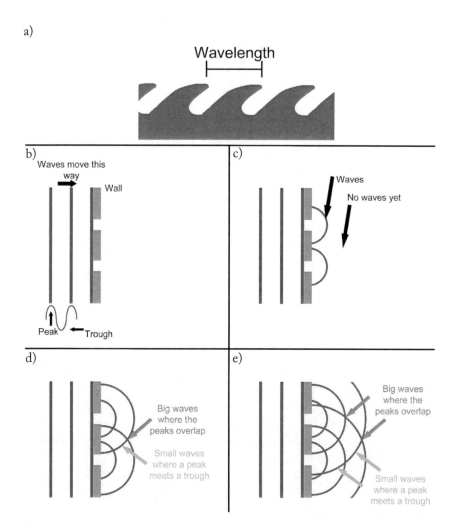

FIGURE 5.1 Water waves make it easy to illustrate what we mean by a wavelength. In the bottom four pictures, we see water waves moving to the right and then pushing through two holes in a wall and creating a wave-pattern on the other side.

How do we know that light is a wave? In order to answer that question, we start by explaining one of the important properties of waves.

Imagine you are sitting at the beach, but this time, the waves are crashing directly into a wall. You will not see any waves on the other side of the wall because the waves cannot travel through it. However, the situation can be very different if there is a single small hole in the wall. As the waves roll in, the water pushes through the hole and produces waves on the other side. If you look at these waves, they will no longer be straight lines, but rather semi-circles—like half of the circle you would see if you dropped a rock in a puddle of water.

Things change even more when there are two holes in the wall, as shown in Figure 5.1b. If the two holes are near each other, we can observe the special property of the waves: instead of continuing through, they spread out and interact with each other, creating a pattern. In some places, the water is rough, and in other places, it is quite calm. While it may look like chaos, it turns out that the pattern is readily understood. Where the peaks of two waves meet (and interfere with each other), the waves are very big. Where the peak of one wave meets the wave trough of the other, the two waves interfere in such a way to cancel each other out, and it is calm.

We see the same patterns with light waves as we do with water waves. But instead of producing the wavy and calm regions you'll find with water, a laser shown through two tiny holes in a wall will instead result in bright and dark spots behind the holes (see Figure 5.2). These spots confirm that light is a wave, and the distance between them allows us to measure the light's wavelength. The light we see with our eyes has a wavelength of about a half of a millionth of a meter (10^{-6} m), which is about ten times smaller than the distance across a red blood cell (see Figure 3.1). No wonder most of us never knew light was a wave!

We learn more about white light by shining it into a prism, as shown in Figure 5.2b. When we do this, the prism breaks the light into different colors of the rainbow: red, orange, yellow, green, blue, indigo, and violet (Roy G Biv). We call this distribution of colors a **spectrum**, and from it, we deduce that "white" is not a color of light at all—it is a blend of all colors. Since we can measure the wavelength, like above, we can show that each color has its own special wavelength. For example, light with a wavelength of 7×10^{-7} m is a nice shade of red.

Devices that look at light from an object and separate the light into colors are called **spectrometers.** You can attach one of these to the end of a telescope and use it to look at the light from the stars, as shown in the figure.

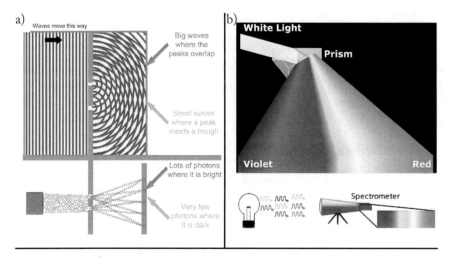

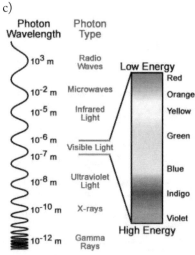

FIGURE 5.2 In the same way that water waves interact as they push through two small holes in a wall, shining laser light (the wiggly lines) through a pair of holes produces the same kind of pattern. In "b," we see white light shining into a prism and separating into different colors; each color has its own wavelength. We now know that we can have light of any wavelength, and we give many of them names. While the visible spectrum is the only part we can see directly with our eyes, it is only a small part of the different types of light that exist. Color versions on Page C-3.

Spectrometers also tell us that our eyes cannot see all the different wavelengths of light. The visible light spectrum (the colors we *can* see with our eyes) contains a specific, and very limited, range of different wavelengths. The full range of light

(Figure 5.2c) includes microwaves, radio waves, infrared light, visible light, ultraviolet light, X-rays, and gamma rays. Each corresponds to a different wavelength. From the picture, we can see that red light has a longer wavelength than violet light. As we go more to the "red side" (longer wavelengths) of the visible light spectrum, we get infrared light; as we move toward the "blue side" (shorter wavelengths), we get ultraviolet light.

5.2 LIGHT AS A PARTICLE

Light is also a particle. Scientists have given it the name **photon**, and it was listed among the fundamental particles in Table 3.1. To understand more about what we mean by a "particle," we go back to Chapter 3, when we talked about an electron as a "fundamental particle" and a proton as a "composite particle."

In many ways, when we call something "particle-like," we are saying that it can be counted. You could say that you got hit by one electron, or one proton, or three quarks.

Another important property of a particle is that it moves in a straight line unless something forces it to change direction. Consider a rock. It can be thought of as particle-like because if it is near the surface of the Earth, the pull of gravity will cause it to fall. Alternatively, if it is far away from any planets or stars, it could travel forever in a straight line until it hits something or otherwise interacts.

An electron moves like a rock traveling in space, but if it happens to be near a proton, it could move in a circle around it because of the electric force between the two (more about this in Chapter 7). The same phenomenon occurs when the attraction of gravity causes the Earth to orbit around the Sun. Billiard balls on a pool table likewise demonstrate this property when they either move across the table in a straight line or are forced to change direction by bumping the side of the table or another ball.

What is some evidence that light is particle-like? Let us get back to the light coming from our laser, as shown in Figure 5.2. As we dim the laser, we notice that the color of the light on the back wall stays the same and the spots remain in the same place. The difference is in the brightness of the spots, which get progressively dimmer. From this, we learn that the laser is sending less light.

If we dim things enough, we could count each particle of light after it leaves the laser, "hits" one of the holes in the wall, changes direction, and strikes the back wall. The light is behaving like a particle. Another example of light acting like a particle is in Figure 5.2b, where each photon enters the spectrometer individually (and spectrometers can count photons). In a very real sense, light hits your eye one photon at a time.

Since this is a lot to take in, let's look at some more examples.

A typical light bulb makes photons that hit our eyes. If it is a dim bulb, there are fewer photons. If there is no light at all, there are no photons, and the room is dark. Darkness is the absence of photons.

When you look at a light bulb, the light often looks white. What you are seeing is all of the different visible wavelength photons hitting your eyes. When you shine this same white light into a prism, it breaks the light into different colors. This is light acting as both a particle and a wave.

Ultimately, then, we say light is both a wave *and* a particle, and can be described as either because it has the properties of both. In many ways, this is the essence of quantum mechanics, which we discuss in greater depth in Chapter 7.

One more thought before we start using light to teach us about the universe.

Light has energy, and that energy depends on the wavelength. If you stand next to a heat lamp, for example, you will warm up. But if you have two identical lamps emitting the exact same number of photons, but one with pure red light, and the other with pure violet light (and assuming your skin absorbs all the light), then you will heat up more quickly with the violet light. For this reason, we say red light has lower energy than violet light. Taking it further, the smallest wavelength light (a gamma ray) has the largest energy, and the largest wavelength light (a radio wave) has the smallest energy.

Even though photons have energy, they do not have mass. For those of us who have taken a physics course, this is counter to our intuition. How can they have energy but no mass?

We think of a thrown rock as having energy, often known as "kinetic energy." The more mass or the more speed it has, the more energy it has (see Box 5.1). A simple bathroom scale can measure how much a person weighs. Scientists can measure the mass of an electron. However, every experiment so far has shown photons have exactly zero mass.

If a rock slows down, it has less energy. How can a photon, which always moves at the speed of light and has no mass, change its energy? What changes is not the photon's mass, but rather its wavelength. Blue light has a shorter wavelength than red light. It is more energetic than red light, but is no faster, nor is it more massive. Ultimately, photons are similar—but not identical—to particles like electrons.

The "Energy of Motion"—Kinetic Energy

The higher the speed of a particle (like an electron) in motion, the more energy it has. Similarly, the more mass the particle has, the more energy it has. For example, think of a car. One can think of its energy as being equivalent to how much damage it inflicts if it hits a tree—the more energy it has, the more damage it will do. A large car, like an SUV, moving at 10 miles per hour has more energy than a compact car, like a Prius, moving at 10 miles per hour; the SUV will cause much more damage to a mighty elm. Similarly, an SUV moving at 60 miles per hour will inflict more damage than an SUV moving at 10 miles per hour because it has more energy. Which will do more damage: a Prius moving at 60 miles per hour or an SUV moving at 10 miles per hour? It all depends on the masses of the Prius and the SUV.

Just in case you were wondering, the answer is actually the Prius because speed is more important than mass. Either way, both are significant.

5.3 THE SPEED OF LIGHT AND LIGHT-YEARS

Scientists can measure the speed of light to incredible precision. In fact, the speed is so well-measured that it is better than our ability to define what we mean by a "meter." Thus, scientists have agreed that the speed of light is exactly 299,792,458 meters per second (usually we say 3×10^8 m/sec) and use that to define what we mean by a meter. This speed is incredibly fast. It is about 186,000 miles per second, or about one foot per nanosecond (10^{-9} seconds). This means that it could go around the Earth about 7.5 times in a second.

While we now know that light constantly travels at this specific speed, scientists did not always know this. It wasn't until the 1860s that James Clerk Maxwell and others could claim some kind of understanding about light and calculate what its speed should be. Further improvements in understanding this concept were made by Einstein, among others, in the early 1900s. We will describe these in Chapter 6. Today, the theory of light—and the special speed and mass it predicts for photons—has been tested extensively.

We can use the speed of light to describe vast distances. Stars and galaxies are so far away that it is often easier to express their distance by using light-years, which is—as the words imply—the distance that light travels in a year. Since light travels at about 3×10^8 meters per second, and there are about 3×10^7 seconds in a year, we find that light travels about 10^{16} meters in a year. In other words, a light-year is about 10^{16} meters. This is an enormous distance compared to the size of the Earth, but only a tiny step in the scheme of the universe, as we saw in Chapter 2. The Sun is about 8 light-minutes away from the Earth, and the next closest star after the Sun—Proxima Centauri—is about 4.2 light-years away.

This also means that the light from Proxima Centauri arriving at our eyes today left that star 4.2 years ago. We are not seeing the star the way it is *now*—we are seeing it as it was *then*. Taking this further, light from a star a billion light-years away has been traveling through space for a billion years and is only now reaching us. To look at those stars today is to see them as they were an extremely long time ago.

Viewing history from the light of these ancient stars is the next-best thing to actual time travel. While it is not really a time machine, it can help us understand what stars and other things in the universe looked like in the past. It is possible that the star that produced the light we now see does not even exist anymore. If the Sun magically stopped shining, we would not know for eight minutes.

5.4 THE DOPPLER EFFECT

It turns out that we can use the wave nature of light to study how stars move. To do this, we need to learn about the Doppler effect, which describes what happens to waves when an object sending off—or emitting—waves is moving.

In the 1840s, Christian Doppler was studying sound and things that make sound. Like water waves and light, sound is a wave with a wavelength and a specific speed.[1] Sounds with a higher pitch have shorter wavelengths and sounds with lower pitch have longer wavelengths. When Doppler listened to how an object sounded when he moved it around the room (scientists sometimes call this the "source of the emitted sound"), he noticed that the pitch changed as the source moved. We now call this the "Doppler effect" of waves.

Everyone has experienced an ambulance come toward him, siren blaring, and then pass by. It makes that distinctive "Eeeeeeeeee-yooooouuuuuuuuuuu" sound. The pitch of the sound is higher as the ambulance comes toward you (the "Eeeeee" part), and

---

[1] There is a nice way to show that sound has a speed. When you yell from a mountaintop into a valley, you will hear an echo. This is the sound wave traveling to the ground and then bouncing back to you. You can also measure how far away a thunderstorm is from you by counting the number of seconds between seeing the lightning and hearing the thunder. The light will hit your eyes almost immediately, but the sound travels at a speed of about one mile per five seconds, so it takes longer to reach your ears. Thus, if you see lightning, and count five seconds before hearing thunder, the storm is about a mile away.

lower as it moves away from you (the "yooooouuuu" part). That is the Doppler effect in action. If you have not ever noticed this, I encourage you to do so. It is quite neat.

To explain how this happens, we look at an ambulance with its siren on (see Figure 5.3). The siren produces sound waves, which is what we hear. When the ambulance is stationary, the sound waves have a specific wavelength that is the same in all directions. When the ambulance starts moving, the sound waves behind the vehicle are "stretched out," and the waves in front of the vehicle are "scrunched up." This causes the wavelengths to change; the ones in the direction of motion are shorter, and the ones going in the opposite direction are longer. Shorter wavelengths mean a higher pitch. To a person in front of the ambulance, the siren, therefore, has a higher pitch; and to a person standing behind it, the siren has a lower pitch. If the ambulance approaches you and then passes you, you get one after the other... the familiar "Eeeeeeeeee-yooooouuuuuuuuuu."

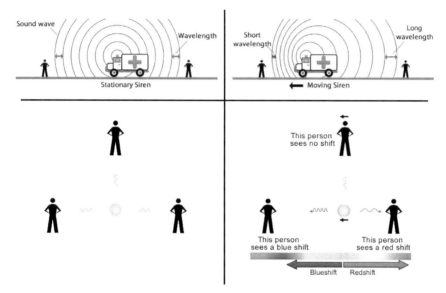

FIGURE 5.3 The Doppler effect on sound and light. In the top left picture, the ambulance is standing still. For this reason, the wavelength of the siren's sound waves is the same in all directions; it sounds the same to both people. The top right picture shows what happens when the ambulance is moving forward. The wavelengths of the sound waves hitting the person in front of the ambulance are scrunched (shorter); the wavelengths hitting the person behind the ambulance are stretched (longer). The person in front hears shorter wavelengths as having a higher pitch, and the person in the back hears the longer wavelengths as having a lower pitch. In the bottom row, we have the same thing, but instead of an ambulance emitting sound, it is a hypothetical star emitting a single color of green light. In this exaggerated example, if the star is moving with a very fast speed, one person will see the light as red (a large red shift), and the other person will see it as blue (a large blue shift). Note that there are no stars that emit only a single color in nature, and the speeds required to create this big of a Doppler shift would be very close to the speed of light.

Since both sound and light are waves, the Doppler effect also occurs for light. Our eyes are not good enough to notice this effect, but with the right spectrometer, you can observe it clearly.

For example, if we look at light from a star or galaxy that is moving quickly away from us (shown in exaggerated form in Figure 5.3), the light coming from it will appear to have a wavelength that is ever-so-slightly "stretched" relative to what we expect. Said differently, the wavelength of the light has shifted toward the red end of the visible light spectrum. We call this a "red shift." The light from stars moving toward us will appear to have shorter wavelength light, so the light is shifted toward the blue end of the visible light spectrum. We call this a "blue shift."

Consider an oversimplified example: a moving star that emits only one color of light. While real stars emit all the different colors of light (our Sun mostly emits yellow), our hypothetical star emits only green light (we will talk about how we know what color of light stars are emitting in Chapters 7 and 8).

In Figure 5.3, we see three different people observing the hypothetical star. If the star is not moving, all three will see green light. Continuing this simple example, if the star is moving at a particular speed, the person in the direction of motion sees the light as blue, the person in the opposite direction sees it as red, and the person moving together with the hypothetical star sees the light as green.

If the star is moving at a slower speed, the first two colors will not be blue or red, but they will be bluer or redder since they have been Doppler-shifted. Note again that this is just a simplified example. While powerful spectrometers can measure the shift, the color of a star does not change very much at all unless the star moves with a speed that is very close to the speed of light.

Next, imagine yourself in an even more complicated scenario, where you are looking at four different versions of our hypothetical green star (see Figure 5.4). In two versions, it is moving quickly, and in two others, it is moving slowly. In addition, two are moving toward you, and two are moving away—there are four possibilities. Since you know the color of light emitted, you can tell from the observed color of light whether the star is coming toward you or moving away from you. Not only that, but you can even determine the speed of the star.

One final thought before we leave this idea. The bottom part of Figure 5.4 shows an observer with two more versions of the same type of star. She observes that both are moving away from her, but that one is moving faster (more red-shifted) than the other. However, the observer cannot tell whether she is at rest and they just have different speeds, or if the two stars are moving at the same speed and she is moving toward one of them. In fact, there is no correct answer. Both are

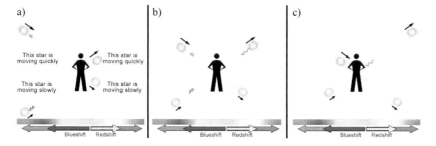

Top Row: An oversimplified view of a hypothetical star that emits only green light and is moving with a speed close to the speed of light. If we know the star emits green light, we can tell from the color of the light hitting us if it is coming toward us or moving away from us. We can also tell how fast it is moving. The three pictures show snapshots in time as the light from the star moves toward our observer. Note that the speeds are greatly exaggerated. (Color version on C-3.)

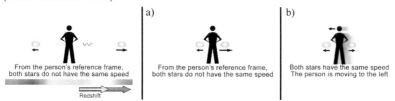

Bottom Row: An observer watches two different versions of our hypothetical star as they move away from her. She is standing still and sees them moving away from her with different speeds. She cannot tell whether the reality is scenario a (middle figure), where she is at rest and the stars have different speeds, or scenario "b" (far right), where she is moving toward one of the stars and both stars have the same speed. Relativity says that the two scenarios are equivalent.

FIGURE 5.4

equivalent, or relative. This is the "relative" in Einstein's theory of relativity, and this scenario will be important in the future. We will come back to it in a later chapter when we talk about stars and galaxies.

5.5 SUMMARY: PUTTING IT ALL TOGETHER

The light we observe from the night sky tells us a lot about the universe. Light reaching us today from distant stars shows us what the universe looked like billions of years ago. From the Doppler effect on light, we can learn about the speed and direction of stars and galaxies. Later chapters will show how we use all of this to tell us what stars are made of and how far away they are. Before that can happen, however, we need to know some more about the great theories of general relativity and quantum mechanics.

Gravity, General Relativity, and Dark Matter

CHAPTER 6

For more than seventy-five years, the theories of general relativity and quantum mechanics have done an incredible job of correctly predicting the outcome of innumerable experiments. General relativity is useful for describing gravity and the large things detailed in Chapter 2. Quantum mechanics describes things that are tiny, such as atoms and molecules—the sizes we saw at the end of Chapter 3.

In this chapter, we will focus on gravity. We will start with a useful—albeit incomplete—version of gravity originally put forth by Newton, who was the first to describe how things are attracted to each other because of mass. In doing so, he made us all realize that a rock falls to the surface of the Earth for the same reason that planets go around the Sun.

We will also discuss the modern theory of gravity, known as general relativity, introduced by Einstein. We will need this more precise and accurate description to properly understand not only gravity and the evidence for dark matter, but also other mysteries like black holes and the expansion of the universe.

We begin by examining some highlights of each theory, continue with the experiments in the early part of the last century that confirmed Einstein's more complete description of gravity, and end with the evidence for dark matter.

6.1 NEWTON AND UNIVERSAL GRAVITATION

In the late 1680s, Newton revolutionized physics by providing a new understanding of how things move.[1] We call his description **classical mechanics**. One of the most important things he noted was that a particle—or any other composite

---

[1] To be fair, many of the ideas were based on the work of others, including Galileo. Of particular importance was the use of mathematics to gain a proper understanding of physical reality. While we are shying away from using mathematics in this text, scientists consider it central to our understanding. They call it the language of nature.

object (like a rock)—always moves in a straight line, and with the same speed, unless there is a **force** to make it slow down, speed up, or change direction.

To help illustrate this idea, we consider some examples.

We start with a rock sliding on the ground. It slows down because of the force of friction. A hockey puck, on the other hand, travels farther because ice has less friction than the ground.

A force can likewise speed things up. Continuing with our puck analogy, a hockey stick can force a puck to accelerate from rest to high speeds. The force of the stick can also change the puck's direction.

A string can exert a force when it is tied around a rock and swung around over your head. The string keeps the rock going in a circle even if the rock stays at the same speed. If you let go, the force of the string will be gone and the rock will fly off. If you are in outer space, it will move in a straight line, keeping the same direction of motion it had at the moment you released it. On Earth this is also true, but it will be simultaneously falling. Another example of this phenomenon is experienced every time the tires on your car push against the road and speed you up, slow you down, or allow you to turn a corner.

Perhaps the most well known of Newton's contributions is in the area of gravity. He described gravity as a force. In his description, *all* things with mass are attracted to each other; it is what makes apples fall to the Earth and the Moon to circle around it[2] (see Figure 6.1). This is a remarkable idea. No matter how big or small the object; no matter how far away things are from each other, they all feel the "pull" of gravity. As far as Newton could tell, this should be true everywhere in the universe. Thus, he named this description **universal gravitation**.

Newton's gravitational theory effectively explains many things, such as why the Moon orbits the Earth and how the Earth orbits the Sun. If there were no gravity, the Moon would simply fly off into space, as shown in the figure. The force of gravity also causes the other planets—as well as things like comets and asteroids—to go around the Sun. And although it's not perfect, this theory is also remarkably accurate at predicting the speed at which these objects move.

---

[2] The Moon is being pulled into the Earth, and that is what makes it go in a circle. This is, in many ways, the same way the rock we mentioned moves above your head in a circle when it is being forced to do so by a string.

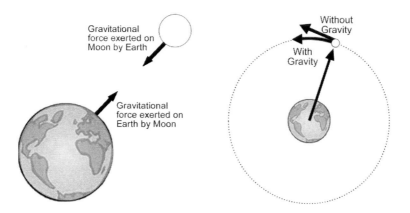

FIGURE 6.1 In Newton's (incomplete) description of gravity, the Earth and Moon attract each other because of the force of gravity between them. If gravity were not present, then the Moon would just fly off into space.

According to Newton:

1. Every object moves in a straight line, with constant speed, unless acted upon by a force.
2. Every object with mass in the universe attracts every other object with mass in the universe.
3. The bigger the masses of the objects, the stronger the attraction between them.
4. The farther the distance between any two objects, the weaker the attraction.

While this theory does an excellent job of predicting how things in nature move, the theory does leave some big questions unanswered: What is doing the "forcing"? Why can't we see what is "pulling" the Moon toward the Earth?

Nevertheless, this model of gravity stood until the twentieth century and is still taught today in high schools and colleges, as it does a worthy job of describing how to build bridges and put a man on the Moon. Then again, it is not good enough to make Global Positioning Systems (GPS) work.

6.2 EINSTEIN AND GENERAL RELATIVITY

While many people learn about Newton and his theory of gravity, most do not realize that our best understanding of gravity actually comes from Einstein.

During the first decade of the 1900s, Einstein began to compose his theory of general relativity and it completely changed the way we view gravity. Instead of gravity being a force that makes things alter their speed or direction as they move through space, general relativity says that what *appears* to be the force of gravity pushing a mass is better described as a mass moving in a "straight line" in what we call "curved space-time." This is complicated, so we will explain more about this particular kind of "straight line" and what we mean by the bizarre-sounding phrase "curved space-time."

Einstein's inspiration for understanding the relationship between space and time came while he was thinking about light. He knew from the experiments of others that light always moves at the speed of light, but wondered what he himself would see if he were moving at different speeds.

Let's use our crime-show theme to explain Einstein's thought process.

What would happen if a detective were driving his car at half the speed of light and turned on its headlights (see Figure 6.2)? The light should move away from the car at the speed of light. Since the speed of light is one foot per nanosecond, in two nanoseconds, the light would be two feet ahead of him.

But what would a second detective, standing on the side of the road, see?

In two nanoseconds, she would see the light move two feet ahead. Because the car moves at half the speed of light, however, it also moves one foot forward during those same two nanoseconds. The second detective would therefore see the photon as only one foot ahead of the car.

Our two detectives get the same answer for the speed of light, but different data for the distance between the light and the car!

This sounds abstract and perhaps unimportant (not to mention it is not clear how we could make a car go that fast), but it has enormous implications. Most people would guess that one detective is right and the other is wrong—although they might disagree on which. Einstein, however, deduced that both perspectives (also like in Figure 5.4) are correct.

To reconcile this apparent contradiction, our detectives must take into account that space and time are more interrelated than most people realize. This becomes

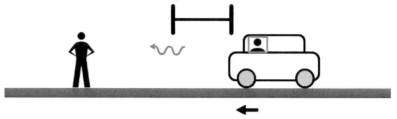

From the perspective of the person on the side of the road, the car moves at half the speed of light and the photon moves at the speed of light. After 2 nanoseconds, the photon is 1 foot ahead of the car.

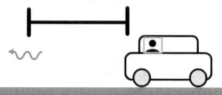

From the perspective of the driver, the car is stationary and the photon moves at the speed of light. After 2 nanoseconds, the photon is 2 feet ahead of the car.

FIGURE 6.2 Two different observers—a person on the side of the road and a car driver—watching a photon move away from a car when its lights are turned on. From the perspective of both people, the speed of light makes the photon travel two feet in two nanoseconds. Since the car moves at half the speed of light, it travels one foot in two nanoseconds. Therefore, from the perspective of the person on the side of the road, the photon is one foot ahead of the car. However, from the perspective of the driver, the car is stationary, but the photon moves two feet ahead of the car in two nanoseconds (since it moved away from the front of the car at the speed of light). Therefore, the driver observes the photon to be two feet in front of the car. The two observers disagree on the result, but relativity explains that both observations are correct.

especially important when speeds are close to that of light. This better understanding is known as **special relativity**. What Einstein figured out is that space (measured with a ruler) is not separate from time (measured with a clock). Rather, both should be thought of as a single combined entity, what we call **space-time**. Space-time has four dimensions: space has three (length, width, and height) and time has one.

If this sounds complicated and bizarre, you are right; it is. You can spend a long time studying Einstein's theory of relativity and there are many good books about it. However, for our story, it's merely important to understand that space and time are intimately related.

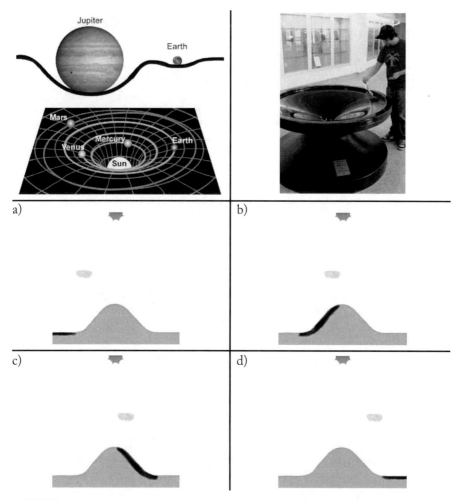

FIGURE 6.3 An artist's conceptions of general relativity. On the top row (top left), heavy masses stretch space-time and cause dents in space-time; an analogy is planets placed on a large, taut rubber sheet. Other masses move in this curved space-time; an analogy is shown in the top right with pennies moving in a gravity well. In the bottom two rows, we show an analogy which illustrates how an object moving in a straight line appears to be moving in a curved path. In these figures, the Sun is shining and a cloud (moving in three dimensions) causes a shadow to be cast on the two-dimensional ground. While the cloud moves in a straight line, the shadow appears to move up and then down the mountain. With these ideas, general relativity describes the inner planets moving in the curved space-time around the Sun, as shown in the top left. (Color versions on Page C-4.)

BIG BANG, BLACK HOLES, NO MATH

Einstein realized that this close space-time relationship has many important implications that can seem quite odd. For example, not only can space and time be bigger or smaller for different people as in the case above, but also they can change or curve. With this understanding, he realized it can describe how two objects with mass attract each other, and how this phenomenon can give us a more fundamental understanding of gravity. It is this theory that we refer to as Einstein's general theory of relativity, or "general relativity," for short.

A basic tenet of general relativity is that space-time can be "warped" by things with mass. An analogy to help you visualize curved space-time, but in two dimensions, is shown in Figure 6.3.

In this figure, we see massive things, like planets, causing depressions on a taut rubber sheet. The line represents space-time and the mass of each object makes it sink into the rubber sheet, creating an indentation all around it. The bigger the mass, the bigger the dent it creates in space-time. While it is not a perfect analogy (some scientists find this analogy potentially misleading for the reasons described in Box 6.1), it gives a good visual analogy for what we mean by the phrase "mass curves space-time."

BOX 6.1

Why Scientists Don't Like the Rubber Sheet Analogy

The analogy of a ball on a rubber sheet is good visually, but really makes many scientists cringe—and for good reason. For starters, with a ball and a sheet here on Earth, the dent comes from the pull of gravity on the ball: there is no force making a dent in space-time. It also implies that a second ball "naturally" rolls to the bottom of the dent in space-time because balls naturally roll down hills. While this is also true, the reason it happens is because gravity is pulling the ball as it goes down the hill. That also isn't fair because there is no force pulling the ball down the hill. A third issue is that in our picture, the dent in space-time only describes things in two dimensions, but the dent in space-time is really in three dimensions. That's harder to draw and visualize, especially for someone just starting to learn the basics. What is important is that the geometry of space-time is curved by a mass in space, and that a second mass can speed up (or slow down) as it moves through this curved space-time, depending on the amount of curvature.

For a more complete and better description, I point interested readers to Kip Thorn's *Black Holes and Time Warps*.

A second picture in Figure 6.3 shows a student dropping a penny into a curved gravity well like those you find in most science museums. The penny's movement around the well gives you a rough idea of how objects move in curved space-time. As the penny moves toward the deepest part of the dent, it will speed up like a ball as it rolls down a hill.

It gets slightly more complicated before we are done, but we need to understand one more thing about general relativity to see how this all ties together. General relativity agrees with Newton's description that an object moves in a straight line in space unless there is a force acting on it. However, in general relativity, an apple falling toward the ground does not feel a force (since there is no force to feel), but it does move in the equivalent of a "straight line" within the big dent in curved space-time created by the enormous mass of the Earth. Specifically, the speed of the apple, in space, increases because the space-time around the massive Earth is more and more scrunched the closer you get to the center of the Earth.

Moving in a "straight line" in curved, four-dimensional space-time can look the same as moving in a curved path in three-dimensional space. This is more easily understood with an analogy since it is hard to think in four dimensions. One of my favorites is to consider how a plane travels from New York to Paris. We tend to think of the plane as traveling directly from one place to the other, but if you pull out a globe and a piece of string, you'll see that the "straight-line" path is actually curved (scientists call this a geodesic).

Another analogy is shown in the bottom rows of Figure 6.3, where we have the shadow of a cloud moving over a mountain. While the cloud moves in a straight line in regular three-dimensional space, the shadow moves up and down the two-dimensional curve of the mountain.

So in summary, Newton's and Einstein's theories state the following:

- Newton: Unless there is a force, things move in a straight line in three dimensions; gravity provides a force.
- Einstein: Unless there is a force, things move in a "straight line" in space-time, which has four dimensions; gravity is not a force, but appears like one because mass curves space-time and objects move along in curved space-time.

It is not hard to draw an analogy for the Earth's movement around the Sun in this "dent-in-space-time" view. While we do not see it this way with our eyes, the Sun creates curved space-time and the Earth moves in a "straight line" in four dimensions within this curved space-time (again see Figure 6.3). The same is true for the other planets. In general relativity, the Earth does not move around the Sun because there is a "force" of gravity, but rather it moves in a "straight line" in the curved space-time created by the Sun.

Don't worry too much if you feel like you are not completely grasping this concept. It is very non-intuitive, and no scientist understood it properly before Einstein. What is important is that, in later chapters, we will be able to use the idea of curved space-time to explain the data we have gathered about our universe over the years. Hopefully, as you read these chapters and see how the ideas are used, it will make more sense. For now, we will look at the great experiment that helped show that general relativity was a better description of nature than was Newton's theory.

6.3 THE GREAT EXPERIMENT

Both Newton's and Einstein's theories aptly describe how apples fall toward the Earth and planets go around the Sun. How do we decide if one is right and the other is wrong? Like a good detective or scientist, we need additional evidence. What experiment can we do where general relativity predicts a different outcome than Newton's theory?

To answer this, we will use perhaps the world's most well-known equation: $E=mc^2$. In this equation (which is only true for things that have mass in general relativity), E is energy, m is mass, and c is the speed of light. This equation therefore means that energy and mass are equivalent to each other as long as we take into account the speed of light.

If we consider light—which has no mass—in Newton's theory, then it should not be forced: no mass, no force of gravity. Specifically, if nothing forces or pushes it, light always travels in a straight line.[3]

General relativity, on the other hand, says that light has energy, which is equivalent to mass, so it should move the same way a mass would move in curved space-time. Therefore general relativity predicts that the path of light traveling through space will bend because of the curvature of space-time—not because of a force on it.

In the late 1910s, experiments were conducted to test which of the two theories correctly predicted what actually happens. We describe a basic version of one such experiment as shown in Figure 6.4.

---

[3] Note that this is a simplified interpretation of Newton's theory. A more thorough reading says that while the light's path will, in reality, bend, it will only do so half as much as general relativity predicts.

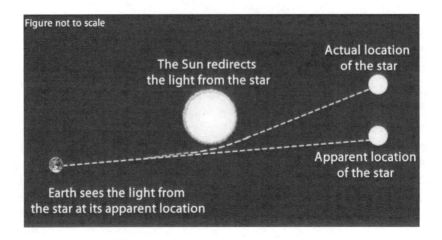

The Sun redirects
the light from the star

Actual location
of the star

Apparent location
of the star

Earth sees the light from
the star at its apparent location

FIGURE 6.4 Light as it travels from a star, past the Sun, and into our eyes here on Earth. In a famous experiment in the late 1910s, scientists looked at light from a star going past the Sun. They did so during an eclipse, to cut out the Sun's glare. The path of the star's light bent and gave the optical illusion of the star being in a different place. This effect is known as gravitational lensing. The difference between the apparent and actual location of the star was exactly consistent with what Einstein predicted and was a stunning confirmation of general relativity.

Astronomers studied the location of the Sun and the stars over time and knew exactly where each should be at a given point during the day. They then focused on a set of stars they knew would be located almost directly "behind" the Sun from their view here on Earth. To be able to see the light from the stars and cut out the glare from the Sun, they did this experiment during a solar eclipse.

Relativity predicted that the light from each star would travel along a path that was "bent" by the Sun's gravity as it passed by the Sun; a star would look as though it were located in a slightly different place than where it should be. You can think of the path of light as passing through a lens, like the one in a pair of glasses, and changing direction. It is important to remember that the amount of this **gravitational lensing** by the Sun would not be much. In fact, the expected deflection from the straight-line path from the star to our eyes is more than thirty times smaller than the eye can discern. However, it was expected to be enough to be measured in very sensitive experiments.

The result was that the apparent position of the stars changed as the light passed from them, through outer space, past the Sun, and to the astronomers on Earth. More importantly, the amount of bending was consistent with the amount predicted by the curved space-time theory of general relativity, and inconsistent with Newton's predictions. The fact that there were many stars that did this, not just one, was especially convincing.

This was a stunning advancement of science and won Einstein international acclaim. General relativity showed that not only did scientists have a more fundamental understanding about how things are attracted to each other in the universe, but also that this attraction was due to how space and time work together in a single space-time.

There are many instances where Newton and Einstein make slightly different predictions about the details of gravity and motion. After decade of experiments in every case, we have found that Newton's predictions were a little bit off, and Einstein's were more accurate. Modern and more powerful tests using gravitational lensing of light have produced even more conclusive findings.

6.4 DARK MATTER IN GALAXIES

As mentioned in Chapter 2, there is clear evidence for dark matter in the universe and that it fills galaxies. Some of the evidence for this understanding comes from two different methods we are now ready to discuss. One method studies the motion of stars in the galaxy due to gravity, and the other uses galaxies to provide gravitational lensing for objects behind them.

We begin with the first method.

The known laws of gravity do a great job of describing how things with large amounts of mass move around each other. For example, it predicts the speed of the planets around the Sun. It likewise gives an intuitive explanation of why the inner planets (Mercury, Venus, and Earth) travel around the Sun with speeds that are much faster than that of the outer planets (Saturn, Uranus, and Neptune). The agreement between the predictions and the observed speeds of the planets is nearly perfect (see Figure 6.5).

In the 1970s, scientist Vera Rubin compared the observed speeds of stars around the center of a galaxy to the predictions of gravity. Since the amount of light from a star is related to its mass, the amount of atomic mass in the center of a galaxy is readily predicted. Other methods allow us to measure the amount of atoms in the disk and far outside the disk.

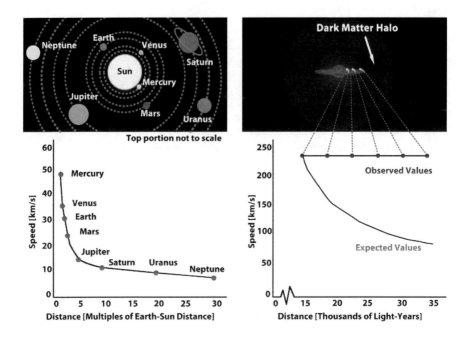

FIGURE 6.5 These two figures show graphs that tell us the speed of a planet or star (vertical axis) and how far away it is from the center of the Sun or galaxy (horizontal axis). For example, from the graph on the left, we see that Neptune is about thirty times farther away from the Sun than is the Earth and it moves about five times slower. We also note the predicted (solid line) and observed speed of the planets (dots) around the Sun. There is remarkable agreement. The figure on the right shows a simplified version of a galaxy. If the galaxy were made of stars alone, then we would see the stars in its outer realm possessing a speed denoted by the solid line indicating "expected values." However, the real data looks more like the observed data plotted by dots on the horizontal line. Our best understanding is that this means that there is a large amount of dark matter and that it is located in a halo surrounding the galaxy.

The left hand side of Figure 6.5 shows a comparison of the observed speeds of the planets to their predicted speeds based on the mass of the Sun. The right hand side shows the expected speed of the stars in the outer part of the galaxy, assuming all the mass of the galaxy is in its stars. In both cases, orbiting objects far from the center are expected to move much more slowly than objects closer to the center.

The data for the stars in a galaxy, however, do not match that prediction at all. Instead of moving like the planets in our Solar System—with stars far away from the center moving at slower speeds—Rubin found that the stars far out on the disk actually move at virtually the same speed as the stars closer in.

While this could be construed as evidence against general relativity,[4] we now understand that the stars are feeling the gravity of a far larger amount of mass (matter) that we cannot see directly. This dark matter—felt, but not seen—is what makes the stars in the outer part of the galaxy rotate quickly around its center. This is where dark matter gets its name.

We can work backward to determine not only how much extra mass is needed to make the stars move this way, but also how this mass would need to be spread around the galaxy to reconcile the data with the predictions. Figure 6.6 shows what the orbits of stars around the center of the galaxy would look like with and without dark matter. Detailed calculations show that we often need about five times as much dark matter as atomic matter in a galaxy, and that dark matter engulfs the galaxy like a giant ball.

We can also use gravitational lensing to "observe" dark matter in a galaxy (and in-between galaxies). As in the optical illusion we saw in Figure 6.4—where the Sun changed the apparent location of a star as the star's light passed by it—we can look at light that gets lensed along its way through the dark matter in a galaxy.

Basically, since there is a large amount of dark matter in a galaxy, its mass bends space-time a great deal, and the curved space-time can significantly bend the path of light. Since the amount of mass is smoothly distributed around a galaxy, any light coming through the dark matter is lensed back toward the middle of the galaxy.

Let us consider light from a distant galaxy; in particular, what happens to light as it passes through dark matter in a nearer galaxy. This is shown in Figure 6.7 and we will refer to them here as the near-galaxy and the far-galaxy. In this case, we can see light from the far-galaxy as it passes by and through the near-galaxy. If the far-galaxy is exactly behind the near-galaxy, then we will not see the far-galaxy directly, but we can see the lensed light from all sides of the far-galaxy in a giant circle, known as an **Einstein ring**, around the near-galaxy. Many of these have been observed in nature.

By using these types of measurements we can determine the quantity and distribution of dark matter in the near-galaxy. These results can be directly compared to measurements of star orbiting speeds in the near-galaxy.

Remarkably, despite how different the speed-of-stars measurement and the lensing-measurement techniques are, they produce similar results for both the amount

---

[4] Many people have tried new versions of gravity, but no one has come up with a theory that explains the other data as well as does general relativity.

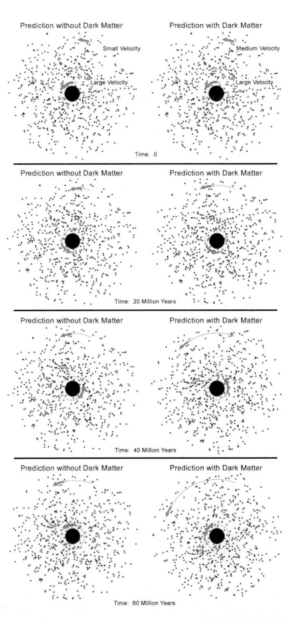

FIGURE 6.6 Two simulations of how fast stars rotate around the center of a galaxy. Notice how the stars near the center are moving quickly. The left side column shows how fast stars should move if there were no dark matter. The right column shows how fast they would move if there were a halo of dark matter in the galaxy. The stars on the outer part of the galaxies are expected to move very differently. The actual data looks much more like the right column than the left.

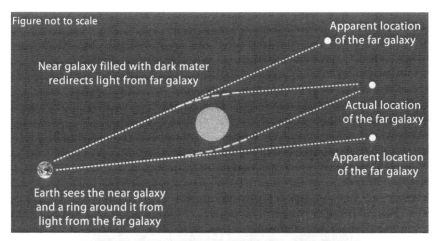

Figure not to scale

Apparent location ● of the far galaxy

Near galaxy filled with dark mater redirects light from far galaxy

Actual location of the far galaxy

Apparent location of the far galaxy

Earth sees the near galaxy and a ring around it from light from the far galaxy

NASA, ESA, A. Bolton (Harvard-Smithsonian CfA) and the SLACS Team

FIGURE 6.7 One way that light from a distant galaxy would look to us here on Earth when there is another galaxy directly in the way. In the special case that one galaxy is directly in front of another, there can be so much mass in the near galaxy that its gravity lenses the light from the galaxy behind it and creates what is known as an Einstein ring. An example with real data is shown in the bottom. (Color version on page C-5.)

of dark matter and how it is distributed. In addition to providing strong evidence for both general relativity and dark matter; they give us confidence that the experiments tell a consistent story. We will come back to dark matter in galaxies in Chapter 15.

Ultimately, general relativity brilliantly describes the space in which we live. It is a central part of the story of how we think the universe has changed over time, from its earliest beginnings until today. We will see this more vividly in Chapter 11 when we discuss how space changes shape during the expansion of the universe.

Now that we have discussed gravity and general relativity, we will tackle quantum mechanics and find out what this theory teaches us about the universe's smaller things.

Atomic Physics and Quantum Mechanics

CHAPTER 7

At this point, we stop to take stock of some of what we have learned, and to say more about where we are going and why.

We have looked at the big things in the universe (Chapter 2) and discussed gravity and general relativity (Chapter 6), which describe how all the big things out there move. We have considered small things like light, electrons, and protons (Chapters 3 and 5), but we have not mentioned how these small things are held together, or how they interact with each other. The time has therefore come to talk about quantum mechanics. As we do so, we will also touch on electromagnetism and atomic physics. We will finish our discussion in the next chapter when we focus on the **strong force**, nuclear physics, and chemistry.

We might summarize Chapter 3 in one sentence: Big things are made of small things—lots of small things (a human body has about 10^{27} atoms).

There are many questions we can ask. For example, "What keeps the electrons in an atom so close to the protons?" The standard answer is that electrons are negatively charged, protons are positively charged, and that opposite charges attract each other.

There is a great deal of truth to this. The attraction between charged things is often referred to as the **electromagnetic force,** or **electromagnetism**. As we will see, the way atoms are held together is more complicated than just the electromagnetic attraction between electrons and protons. To describe atoms and other small things, we need quantum mechanics (addressed in this chapter) to understand how all small things interact. We also need to recognize how other forces, like the "strong force" (described in the next chapter), work.

As we learn about quantum mechanics, we will discover that the special way it keeps atoms together also makes them produce photons with certain colors/wavelengths. In some ways, these special wavelengths are like fingerprints. As detectives, we can use **atomic fingerprinting** to provide evidence that stars are made of atoms and, for reasons we will discuss in the next chapter, which kinds of atoms. The fingerprints will also come in handy to help determine both the direction that the stars move, as well as how fast they are going. Ultimately, the more we know about the stars, the more we can learn about the universe itself.

We are now ready to look at atoms and discover how they produce the light we see from stars in the night sky.

7.1 The Fundamental Building Blocks of Nature

In the seventeenth and eighteenth centuries, physicists and chemists studied matter in its different forms: solid, liquid, and gas. Each seemed to have things in common. For example, gases or liquids could be combined to make other gases or liquids. Scientists hypothesized that matter is made up of different types of atoms, that atoms make up molecules, and that these molecules interact with each other to produce bigger molecules. Lots of molecules interacting together are called either a gas, liquid, or solid; which one the molecules create depends on the molecule type and on such factors as temperature.

A good example of temperature affecting the outcome of molecule interaction can be seen with water, which is made up of hydrogen and oxygen atoms put together. When water is hot, it is the gas we call steam. At room temperature, it is the liquid we call water. When it is cold enough (below freezing), it is the solid we call ice.

By the early 1900s, there was ample evidence that matter was made of atoms, but researchers did not know what atoms looked like, what kept them together, or how they worked.

Figuring out what is inside an atom is not easy. Remember—we now know that atoms are about 10^{-9} m across (about ten million lined up next to each other across the width of your finger), but scientists did not have a way of studying such small things back then. How did they figure it out?

We can describe the experiment researchers used with an analogy (see Figure 7.1). Let us say we have a sealed bag or want to figure out a bag's contents without opening it. One rather unusual way to accomplish this would be to take a gun and shoot lots of bullets at the bag, and then see what happens to the bullets after they have flown through the bag and have come out the other side.

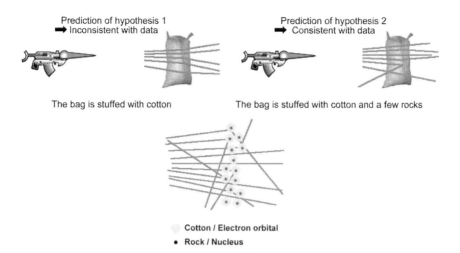

Prediction of hypothesis 1
➡ Inconsistent with data

Prediction of hypothesis 2
➡ Consistent with data

The bag is stuffed with cotton

The bag is stuffed with cotton and a few rocks

  Cotton / Electron orbital
• Rock / Nucleus

FIGURE 7.1 A machine gun shooting bullets at a bag stuffed with cotton and rocks as an analogy of an early experiment firing helium atoms at a thin gold target that helped determine the nature of matter. Experiments showed that the structure of matter is much more like the figure on the top right, and not like the one on the top left.

By studying the bullet holes on the wall behind the bag, we would get more information about its contents. One possibility is that the bag is filled with cotton. In this case, the bullets will go almost straight through the bag and make holes in the wall almost directly behind it.

If we consider the experiment from the perspective of studying the bullet holes in the wall, then note that virtually all the bullets ended up directly behind the bag, we conclude that there is not much in it, or it is filled with something soft, like cotton.

Another possibility is that the bag is filled with heavy rocks. In that case, the bullets might bounce off the rocks, and none would hit the wall behind the bag.

A third possibility is that the bag contains mostly cotton, but with a few rocks mixed in. In this case, some of the bullets might not scatter very much (only hit cotton), some might scatter a little bit (a glancing blow off a rock), and some might scatter a lot (hit a rock dead-on). Obviously, there are many possibilities.

Instead of bullets and a bag, a scientific team led by Ernest Rutherford in the early 1910s experimented by shooting one type of atom at another type of atom. Specifically, they shot helium atoms at a thin strip of gold. What they observed is shown in the bottom part of Figure 7.1.

For most helium atoms, there was very little scattering, but a few helium atoms scattered a great deal. Rutherford had no idea that almost all the mass of the atom was located at its center, so he was shocked by this outcome. He is reported to have said, "It was as if you fired a 15-inch shell at a piece of tissue paper, and it came back and hit you."

Today, we are not surprised by this outcome. We understand that this occurred because our world is put together with the structure illustrated in Figure 3.2. Most of the space in any atom is empty, with a little bit of hard and heavy stuff (the nucleus) and a little bit of soft stuff (electrons) thrown in.

In Rutherford's experiment, then, most helium atoms did not scatter very much because the gold strip was composed of gold atoms, which were chiefly empty space between the electrons and the nucleus. Also, each electron swirling around the gold nucleus is eight thousand times lighter than a helium atom. Except for the rare occasion when the helium atom hit the nucleus, it was like a bullet passing through cotton. However, since gold atoms are fifty times heavier than helium atoms, when the helium atom did hit the gold nucleus, it scattered like a bullet striking a small brick. This didn't happen very often, but when it did, it created a powerful ricochet.

After decades of this type of detective work, scientists settled on the following understanding of the makeup of matter: different types of atoms (hydrogen, helium, carbon, etc.) are varying kinds of nuclei surrounded by electrons. Many years later, scientists discovered that the nucleus is actually comprised of smaller components—neutrons and protons. With these few particle types alone, all the atoms in the universe are constructed. We will learn more about the nucleus in the next chapter. For now, we focus on what holds the electrons to the nucleus.

7.2 ELECTROMAGNETISM: IT'S WHAT HOLDS ATOMS TOGETHER

Why do the electrons stay near the protons and form atoms? The simple answer is because of their electric charge. Protons have been observed to have a positive electric charge, electrons have been observed to have a negative electric charge, and opposite charges are observed to always attract each other. The theory that governs the interactions between electric charges is called electromagnetism. It is summarized in Figure 7.2.

Opposite charges attract each other   Like charges repel each other

Neutral particles neither attract nor repel

FIGURE 7.2 The attraction of charged objects in electromagnetism.

There are many similarities between electromagnetism and gravity. For example, with gravity, every object in the universe pulls on every other object in the universe. With electromagnetism, every charged object in the universe pulls on every other charged object in the universe. With gravity, the bigger the mass, the bigger the pull. With electromagnetism, the bigger the charge, the bigger the pull. In both cases, the farther the distance between the objects, the smaller the pull.

But gravity and electromagnetism differ in important ways. For instance, with gravity, all massive objects are attracted to each other (as far as we know). However, in electromagnetism, there are three types of charges: positive, negative, and neutral. While opposite-sign charges attract each other, like-sign charges repel each other (see Figure 7.2). Neutral particles are neither attracted nor repelled by things with positive or negative charges. A positive charge and a negative charge can be combined to become neutral, but charge cannot be created or destroyed.

Finally, electromagnetism is *much* stronger than gravity. The repulsion between the negative charges of the two electrons is so powerful that it is 10^{42} times stronger than the attraction of the masses of the two due to gravity. Just to remind you: a million is 10^6, a billion is 10^9, a trillion is 10^{12}—electromagnetism is 10^{42} times stronger! Electromagnetism is so much more powerful that the extra attraction due to gravity is essentially irrelevant inside atoms. We will talk more about what keeps protons together despite their repulsion in the next chapter, but for a quick answer, see Box 7.1.

With this simple understanding about the similarities and differences between gravity and electromagnetism, we can describe an atom using a model. The simplest atom is a hydrogen atom (see Figure 7.3). It consists of one proton and one

If like charges repel, then why don't the two protons in a helium nucleus force the nucleus to come apart? Similarly, why don't the two same-charge quarks in a proton force it to come apart? The answer is that there is another force that we will discuss in Chapter 8 that holds protons together: the strong force. As we will learn, the strong force between two quarks inside a proton is about a hundred times stronger than the electromagnetic force. This means that you can virtually ignore electromagnetism inside protons.

electron and they are attracted by their charge. A good—but ultimately wrong—first analogy for the attraction between the electron and the proton is the attraction of the Moon to the Earth. Since the Earth is so much more massive than the Moon, the attraction of the Earth makes the Moon orbit around it. In the same way, since the electron is so much less massive than the proton, the electromagnetic attraction makes the electron orbit the proton.

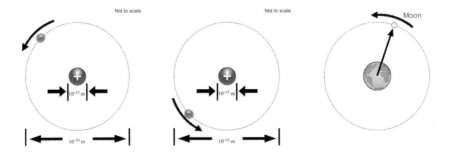

FIGURE 7.3 In the simple (wrong) model of the atom, the negatively charged electron orbits the positively charged proton in the same way the Moon orbits the Earth.

This orbiting-like model is simple, but, unfortunately, it does not fully describe atoms in the real world. If the part of electromagnetism that is similar to gravity is correct, then an electron going around the nucleus should be able to orbit at any distance from the nucleus. Gravity predicts, and we observe, that the planets and objects like comets can be at different distances from the Sun. But this is not even close to what happens for atoms. Experiments show that electrons only orbit at specific distances from their nucleus.

BIG BANG, BLACK HOLES, NO MATH

Another problem is that electromagnetism predicts that any charged object, like an electron, should lose energy as it moves around a nucleus, eventually spiraling into the nucleus. And, in fact, this is what we observe in experiments where electrons move in big circles. Thus, we would expect that an electron moving in a small circle around an atom would also spiral inward until it hits the nucleus.

Based on this observation, scientists naïvely calculated that if an electron started orbiting a proton at a distance of about 10^{-9} m (a typical distance we observe electrons to orbit in atoms), then it should take about ten picoseconds (10^{-11} seconds) for it to spiral in, causing the atom to collapse. However, experiments always show an entirely different outcome: An electron can remain at the same distance from the nucleus indefinitely. Many atoms are therefore stable—they keep orbiting forever.

7.3 QUANTUM MECHANICS

For years, scientists tried to come up with simple explanations for why and how electrons orbit in atoms. Could electrons behave like charged particles when they travel in big paths, but more like planets when they move in small circles? When simple ideas like this did not work, scientists finally gave up and tried new and more bizarre theories.

To explain why electrons "orbit" only at certain distances from the nucleus and how atoms keep from collapsing, scientists developed quantum mechanics in the 1910s and 1920s.

There is a lot to know about quantum mechanics—physics students spend many years learning about it—but here we will present just a few points to help you grasp the big picture. If you find yourself troubled by how complicated or bizarre quantum mechanics sounds, you are in good company. Niels Bohr, one of the founders of quantum mechanics, allegedly said, "Anyone who is not shocked by quantum theory has not understood it."

Despite his important contributions to the theory, Einstein hated quantum mechanics and only begrudgingly accepted it. More than seventy-five years later, though, it is still with us, as strong as ever, having passed every experimental test we have given it.

At the heart of quantum theory is the idea that all particles—not just photons—have properties of waves and particles. Electrons have wave-like properties. In the same way that high-energy photons have small wavelengths and low-energy photons have large wavelengths, high-energy electrons have smaller wavelengths than low-energy electrons. When we talk about a low-energy electron, we mean it has a small amount of energy of motion; it has a small kinetic energy. We need to keep in mind this wave-like nature of an electron when we describe the way it moves around a nucleus.

Using quantum mechanics to properly describe how an electron moves around a proton is a fairly sophisticated mathematical procedure. To simplify, we merely say that the predictions of quantum mechanics indicate that since an electron really is a wave, then the path of the electron "wave" wraps completely around a nucleus.

The circumference of the electron's orbit (the distance around the circle) needs to be an exact number of wavelengths (see Figure 7.4). Any positive, whole number of wavelengths—like 1, 2, 3, 4, etc.—will do, but no other values will work. Since it is unusual that only certain numbers work, we use the word "quantum," defined as "a particular amount," for our purposes. In other words, the number of wavelengths it takes an electron to go around a proton is "quantized." This is the "quantum" in quantum mechanics.

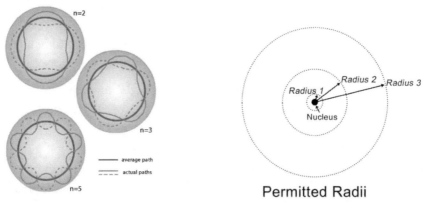

FIGURE 7.4 The quantized "wave-like" orbits of an electron around a nucleus in quantum mechanics. The plot on the left also shows a "smeared"-out version. This indicates that the wave-like nature of the electron is not at any one place at a time. The right side shows the first three different radii that the electron can have. An electron orbiting with radius 1 will have a smaller overall energy than the one at radius 3.

While quantum mechanics makes very clear predictions about wavelength-path combinations, it is worth explaining a little more about what is happening. To do this, we go back to water waves hitting a beach.

In Figure 5.1, waves interfered with each other after they passed through holes in a wall and created both wavy and calm regions. Since an electron wave is neither the peak nor the trough, you can think of it as interfering with itself as it goes around the proton.

Quantum mechanics tells us that the electron is typically "located" where it is really wavy, and never where it isn't wavy at all. Visually, what we call an electron can be thought of as being spread out as indicated by the smeared region in Figure 7.4. Essentially, the electron's wave-like nature means that it is never at a single place at a single moment in time. In some sense, it is at all of the allowable places at once. In fact, the very suggestion of an electron moving along the "circumference" of a nucleus is just an approximation of what quantum mechanics really says. The theory does not actually consider the electron to travel along a single orbital path at all, but indicates that it is always somewhere along the wavelength-path. For more detail, see Box 7.2.

BOX 7.2

It is important to understand that even our complicated-sounding description about the wave nature of electrons moving around a nucleus is itself a simplification. While many people have used the term "orbital" to describe the places that an electron can go, in reality, an electron is never at a single place at a time. The allowed travel area of an electron is more like the surface of a balloon than like a structured path (see Figure 7.4). Scientists, who have spent decades dotting the i's and crossing the t's on the quantum theory (and testing it experimentally), would say that the surface shown in the picture describes the probability of where the electron could be measured at any given time.

Each allowed wavelength-path combination dictates a great deal about where the electron can be in the atom, as well as the electron's energy. For example, if only specific wavelengths are allowed, only electrons with specific energies can exist in an atom. So if an electron has that special energy, it can only exist at a certain distance from the nucleus.

Lower-energy electrons have longer wavelengths, so it takes fewer wavelengths to go around the nucleus. Higher energy electrons have shorter wavelengths, thus it takes a greater number of wavelengths to go around the nucleus. Equally important is the fact that when an electron moves closer to a proton, it gains energy in the same way that a rock gains energy as it falls to the surface of the Earth. We call this **potential energy**. The total energy of an electron is the combination of both the kinetic energy and the potential energy.

In principle, an electron in an atom can have any one of the allowed energies. Each allowed energy is called an **energy state** or **energy level**. Electrons with low energy are in a low-energy state, while electrons with high energy are in a high-energy state. Being completely out of the atom is also an allowed energy state.

Electrons can move from one energy state to another if they gain or lose energy (we will talk about how this happens in the next section). In the same way it takes energy to pick up a heavy rock from the ground and lift it over your head, it takes energy to raise an electron from a low-energy state (small radius) to a high-energy state (large radius). Since electrons can only exist in an atom at special radii (or out of the atom), they cannot spiral down into the nucleus because the intermediate radii the electron would spiral through are not allowed. This is the reason why atoms are stable. For more on this, see Box 7.3.

BOX 7.3

We often talk about quantum mechanics only affecting sizes of a molecule or smaller, but in reality, it impacts things of all sizes. While it is amazing that electrons orbit around a nucleus in quantized positions, quantum mechanics also plays an important role when a planet orbits the Sun—even if that role isn't as noticeable. If you like, you can think of the Earth moving around the Sun as being in a quantized state, where its number of wavelengths is equal to one zillion (a fake, but really big number). In this case, moving to a distance from the Sun in the one zillion-minus-one-state is not noticeable.

It is worth saying a few words about the phrase "not allowed." Let us go back to the speed of light.

As we have seen, moving faster than the speed of light is not allowed in nature. I do not have a good reason *why* nature has chosen (if we can say nature chooses) to make the speed of light the fastest speed allowed, but we observe it to be true in all known experiments. Moreover, theories that are based on this observed fact make further predictions that are also borne out in experiments.

Theories that allow things to move faster than the speed of light have no experimental evidence supporting them. In the same vein, we do not really have a good reason why nature has "chosen" the world to be quantum mechanical. What we do know is that all the experiments we have done on the universe give answers that are consistent with the laws of quantum mechanics. In nature, and as predicted by quantum mechanics, spiraling does not occur, and thus atoms are stable.

One final thought before we move on to electrons and light: While it might seem counterintuitive, it turns out that the lower the total energy of the electron, the closer it is to the nucleus; the higher the total energy of the electron, the farther it is from the nucleus. This is similar to a coin as it falls down a gravity well. While it speeds up as it goes around, it is actually losing energy due to friction and sound. As we will discover, the same is true with atoms. The electron has a higher speed (smaller wavelength) when it is closer to the proton, but it had to lose energy to get there. Since there is no friction, there has to be another explanation.

7.4 ELECTRONS, LIGHT, AND HOW THEY INTERACT

Atoms that interact with light can gain or lose energy when they do so. Since atoms are composite particles, what is really happening is that the photons are interacting a bit with the protons within the atom, but primarily with the atom's electrons. The way electrons give off (emit) or capture (absorb) photons dictates how atoms interact with light. For this reason, we will take a short digression to talk more about how electrons and photons interact with each other. When we are done, we will come back to the interaction of photons and atoms.

The simplest type of interaction between electrons and photons is like balls bouncing off each other on a pool table if we ignore the friction that eventually makes them slow down and come to a stop. In the same way that a ball at rest on a pool table becomes more energetic when it is struck by the cue ball (it speeds up), electrons can become more energetic when they are bumped by photons. Similarly, when the cue ball interacts, it loses energy (slows down). The amount of energy the pair of balls has before the collision is the same as the energy the pair has after the collision. This is known as **conservation of energy**. The amount of energy each particle has can change, but the *total* amount for the pair has to stay the same. So an electron can gain or lose energy when it interacts with a photon, and vice versa for the photon.

The interactions among small particles in quantum mechanics are more complicated because of their wave-like nature, but energy is still conserved. In general relativity, it is better to think of the "force" of gravity as mass moving in curved space-time.

How do we better describe electromagnetic force and quantum mechanics? A good analogy is to think of interactions of charged particles as the "emission" and "absorption" of photons (see Figure 7.5). In the figure, the electron and photon approach each other, and then the electron absorbs the photon, which causes it to have a higher energy. The electron moves off into space.

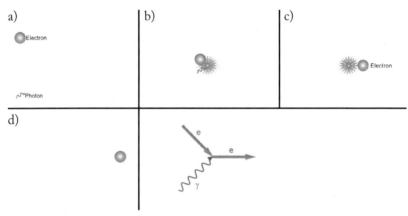

FIGURE 7.5 Some action shots showing the interaction between an electron and a photon in quantum mechanics. In the top set, an electron absorbs a photon and turns into a higher-energy electron. This entire process is depicted in the single Feynman diagram at the end. Think of things as moving toward the right as time goes by: the blue squiggly line with the Greek letter γ next to it represents the path of the photon as it travels toward the electron, which is represented by the solid red arrow with the e next to it. After the two meet, the electron absorbs the photon and moves off into space as a higher-energy electron (also denoted by a red line). The bottom set of plots shows a similar sequence, but here we have a high-energy electron emitting a photon. They both go off into space with a lower energy.

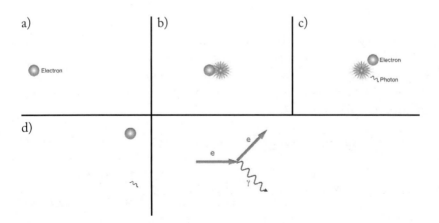

A similar sequence is shown in the bottom part of Figure 7.5, where the electron emits a photon, becomes a lower-energy electron, and both the photon and electron go off on their separate ways. Note that in both cases, the total amount of charge before and after the interaction is the same because the photon is neutral. Also, remember that this is a quantum mechanics thing that only occurs between individual, really small particles—our naked eyes cannot see them because the photons never make it to our eyes. These interactions are represented by what we call **Feynman diagrams**, named after physicist Richard Feynman.

One last thought: A particle can actually disappear, and new particles are literally created. The photon was not "hiding" inside the electron waiting to be released; it, in fact, came into existence out of the energy that the electron formerly possessed. This type of interaction between a photon and an electron can also describe the "force" between two electrically charged objects.

If this concept seems a bit crazy to you, you're not alone. Perhaps it is easier to understand by thinking of this repelling process as two charged particles—like electrons—constantly passing photons back and forth between them.[1] They are not exactly the same type of photons that hit our eyes, but they are close. Scientists call them "virtual" photons.

To try and visualize this, we use a simplified analogy where only one photon is "exchanged." Figure 7.6 shows two electrons placed near each other. One emits a photon and moves off into space (like when you shoot a gun and lurch back). The other absorbs the photon and moves off into space (like if an aluminum can gets hit with the bullet and falls backward). The net result is that the electrons have repelled each other just like electromagnetism says they do. This interaction is the "force" between them; it is what pushes them apart. Physicists say that the force, or interaction, is "carried" by the photon.

Another way of describing this process is that the interaction is "mediated" by the photon in the same way that a business negotiation between two companies can be mediated by an outside party.

The process of attraction between negatively charged objects—like electrons—and positively charged objects—like protons—works much like the repelling process between two particles possessing the same charge. In the case of attraction, though, the photon does not push the particles apart, but rather acts as a mediator that "tells" the electron and proton to move toward each other.

---

[1] If you have taken a course on electromagnetism, it is this passing of photons back and forth that is what we call the **electric field**.

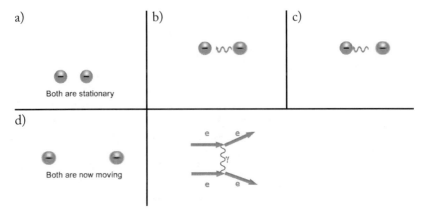

FIGURE 7.6 Virtual photon exchange between two charged particles. In the top set of plots, we have two electrons (same charge) that have been placed near each other. One emits a virtual photon and moves off into space (like shooting a gun and falling backward from the recoil). The other electron absorbs the virtual photon and moves off into space (as if that same bullet hit an aluminum can). The net result is that the electrons have repelled each other just like they should. There is a "force" between them. The Feynman diagram for this appears to the right of the action sequence and shows two electrons *e* as they pass a virtual photon γ between them and repel. The bottom set of action shots shows a virtual photon exchange between a proton and an electron. It is very similar to the top set of plots, but in this case, the virtual photon "tells" the electron to be attracted to the proton and the two particles move toward each other. Again, after the action sequence is the Feynman diagram that shows the proton *p* and the electron *e* interacting by the exchange of a virtual photon γ. Here, the two particles have passed information to each other and created an attraction. This is essentially what is going on inside an atom.

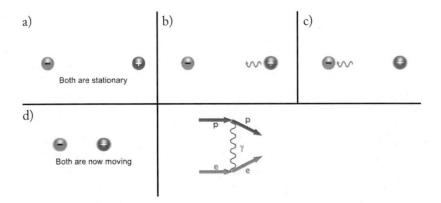

On the bottom of Figure 7.6, we see the proton on the right "emit" a virtual photon. In some sense, this is how it tells the world it is positively charged. The photon then interacts with the electron, which is equivalent to the electron "realizing" that there is a positively charged particle nearby. The electron takes in information, which tells it to approach the proton. We can think of this process as a reason why the photons are called "virtual": they do not really do any pushing, but are essentially just carrying information on how each particle should move.[2] Physicists use Feynman diagrams to describe how the particles "talk" to each other, as shown in the figure.

Now that we know more about quantum mechanics and how various particles can interact, we are ready to talk about atoms again and the evidence for quantum mechanics.

7.5 ATOMS AND LIGHT

Atoms can gain or lose energy based on the way electrons interact with photons. Suppose, for instance, that an electron is moving around a proton in a high-energy state. If the electron is allowed to exist with a lower-energy state, then the electron can transition from the allowed high-energy state to the allowed low-energy state. However, since energy cannot be created or destroyed (again, conservation of energy), this energy has to go somewhere. To accomplish this, the electron in the high-energy state emits a photon (that travels off into space), and the electron moves into the lower-energy state. As we noted before, electrons in higher-energy states orbit farther from the nucleus. Thus, when an electron in a higher-energy state emits a photon and moves to a lower-energy state, it settles into a lower radius orbit (see Figure 7.7).

While the transition of an electron from a high-energy state to a low-energy state happens instantaneously as the photon leaves, you can visualize it as a coin that has moved from a rolling state near the top of a gravity well to one much closer to the center. As the coin nears the center, it also moves closer to the ground. Some of the energy goes to speed up the coin, and some of the energy exits the well in the form of noise.

Now let us return to the atom. Like the coin in the gravity well, when the electron goes from the farther-from-the-nucleus position to the closer-to-the-nucleus

---

[2] It is important to note, especially for those of you who know how balls typically interact when they collide, that this analogy is not a perfect analogy because it does not conserve momentum. Remember, this is not a real photon getting passed back and forth, but information propagating through space at the speed of light from one charged particle to another. Again, virtual photons are the electric field as it permeates space.

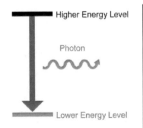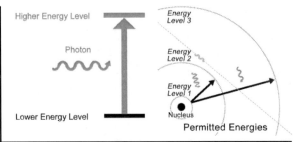

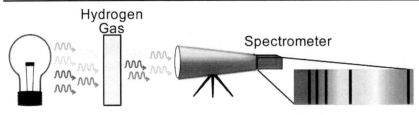

FIGURE 7.7 In the top left plot, an electron moves from possessing a high energy (being in a high-energy state or level) to having a lower energy (low-energy state) by emitting a photon. The inverse is shown next to it. On the top right, a photon can hit an atom and make an electron jump from a low-energy level (small radius) to a high-energy level (larger radius). If the photon does not have the right energy (like the photon traveling along the dashed line in the figure), it can pass through the atom unabsorbed. On the bottom are the spectrometer results for photons from a white light bulb moving through hydrogen gas. Only light of certain colors are absorbed by the gas, so we do not see them. The absent light colors are called spectral lines.

position, it simultaneously speeds up and loses energy that is carried away by the photon. In other words, while it is going a little faster now, it still has less total energy.

While this part of the gravity well analogy works satisfactorily, ultimately it breaks down because the penny can be at any distance from the center, but electrons in atoms can only be at certain distances from the nucleus. Similarly, in the analogy of lifting rocks, while we can lift and lower rocks as we please using our arms or forklifts, with electrons in atoms, typically photons are emitted or absorbed with specific amounts of energy.

The fact that an electron can only change energy levels by absorbing or emitting a photon has enormous implications. If the electron can exist solely in an atom with specific energies, changing from one allowed energy level to a different allowed energy level uniquely determines the energy of the photon involved. This means that only certain energy photons can come out of an atom; the energies are quantized. Similarly, only certain energy photons can be absorbed.

There is no other theory that predicts this type of behavior—a behavior that is regularly seen in experiments. In some sense, this is part of the compelling evidence that our quantum mechanical description of nature is correct. Moreover, it is not just that quantum mechanics predicts that there is emission and absorption of photons; it also correctly predicts the wavelengths of the photons with remarkable precision.

In the analogy of our detective, it is as if our theory has not only told us who the killer is, but also from where in the room he shot the bullet, as well as the exact time the gun was fired. Then, using precision surveillance video footage, a frame-by-frame comparison shows that the data agrees with the theory. By watching the clock in the video, we see that the gun-firing time matches the prediction to the second.

In summary, atoms can interact with the photons they encounter in many interesting, important, and useful ways. A photon can hit an electron in a low-energy level and make it "jump" to a higher energy level within an atom (again see Figure 7.7). Different energy photons can make the electron move into any one of the available energy levels, or even knock it completely out of the atom. However, if the photon does not have the right amount of energy to make it jump into an allowed energy level, then the electron in the atom can ignore the photon completely. The photon will pass by the atom unabsorbed, or bump it like a billiard ball.

Figure 7.8 shows a time sequence of an atom changing energy levels and emitting or absorbing photons. Essentially, an atom can interact with an incoming photon in one of three ways, depending on the photon energy:

1. Low-energy photons either pass right through the atom, or slightly alter the path of the atom by bumping it.
2. If the atom encounters a photon with just the right amount of energy, an electron can absorb the photon and get "excited," meaning that the electron in the atom (or the atom itself, depending on how you want to look at it) goes into a higher-energy state.
3. If a really energetic photon comes along, it can completely knock an electron out of orbit around the nucleus.

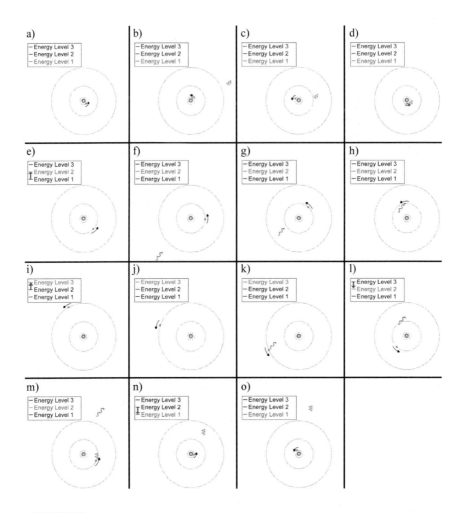

FIGURE 7.8 A time sequence of an atom absorbing and emitting a number of different photons. The atom starts with its electron in the lowest energy state (Energy Level 1). A photon comes in and is absorbed, putting the electron into the second energy level (Energy Level 2). A second photon then comes in, is absorbed, and the electron moves into the third energy level (Energy Level 3). After moving for awhile, the electron emits a photon, which travels off into space while the electron moves down to the second energy level. It then emits another photon and returns to the lowest energy state. There are lots of other types of interactions that can happen. For example, if the electron were in the third energy level, it could just as well have emitted a higher-energy photon and moved to the lowest energy level.

An analogy for these possible outcomes is to think about stuff from space banging into the Moon. If a small meteorite hits the Moon, not much happens. However, a large meteor can create a big crater on the Moon. Finally, a ridiculously immense asteroid could knock the Moon out of its orbit around the Earth and send it off into space forever, never to return.

The special ways that atoms interact with photons are especially useful to us as detectives. They allow us to uniquely identify an atom using what we referred to earlier as atomic fingerprinting.

If we shine white light at a spectrometer (as in Chapter 5), we will see all the colors. But if we insert a bunch of hydrogen atoms between the light and the spectrometer (see Figure 7.7), the photons with the right energies can be absorbed by the hydrogen atoms. The photons that are absorbed do not make it to the spectrometer, so these colors will not show up in the spectrum. Essentially, they will appear as dark lines in the spectrum, known as **spectral lines**. Using quantum mechanics, we can calculate where the spectral lines should be, then observe the accuracy of our predictions in experiments. Spectral lines provide powerful evidence for quantum mechanics, and are incredibly useful when we look at stars.

With some of the basics of quantum mechanics covered, and knowledge of how atoms emit and absorb photons under our belt, we next discuss the different types of atoms. Chapter 8 describes the aspects of nuclear physics and chemistry that help us understand what different atom types look like. This will help us learn about the stars by studying their light.

Nuclear Physics and Chemistry

In the last chapter, we discussed electromagnetism and quantum mechanics, and how electrons are contained in atoms. We also talked about the different colors of light that can be absorbed or produced by a hydrogen atom. In this chapter, we will learn some nuclear physics and chemistry as we move to the bigger and more complicated atoms. From this, we will see that each different type of atom absorbs and produces photons with a unique set of wavelengths. This extends our method of atomic fingerprinting, which will be particularly useful when we look at the stars.

In addition to electrons and protons, atoms can also contain neutrons. Electromagnetism tells us that electrons are attracted to a nucleus because of its positive charge. If this is the case, why don't the protons in the nucleus, which share the same charge as the nucleus, repel each other and break the nucleus apart? Similarly, why do the neutrons, which are neutral, stay in the nucleus?

The answer lies in another force called the **strong force**. As we saw in Chapter 3, both protons and neutrons are composite particles made up of quarks that attract each other and any other quarks nearby. Because of the strong force, the attraction of the quarks inside two nearby protons is greater than their repulsion due to the charge of electromagnetism. It is like the attraction of the smell of fresh coffee in the morning that overcomes the desire to stay in bed all day. Or the reason you still live with your stinky roommate because he pays half the rent.

Ultimately, it is the overwhelming attraction among quarks that keeps protons and neutrons together in different combinations to form different types of nuclei. From there, in the same way that electrons are attracted to protons, we can combine different numbers of electrons with different types of nuclei to form different kinds of atoms.

With this understanding, we have one of the last pieces of the puzzle we need: different atoms produce the distinctive types of light that we see. We can therefore utilize this light to identify atoms in stars.

8.1 Protons, Neutrons, Quarks, and Electrons, and the Atoms They Create

Atoms are composed of electrons surrounding a nucleus (see Figure 8.1). Part of what keeps atoms together is electromagnetism: negatively charged electrons are attracted to positively charged protons in the nucleus (Chapter 7).

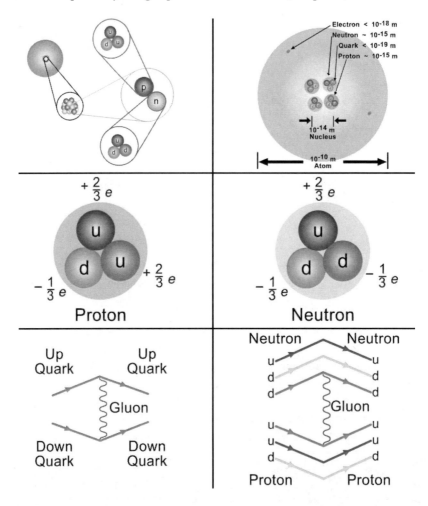

FIGURE 8.1 Zooming in to look at the inner workings of atoms, nuclei, protons, and neutrons with artist's conceptions and Feynman diagrams. Note that none of the pictures are drawn to scale and since quarks and electrons are both wave-like and particle-like, it is not possible to really draw their locations or the paths they take. The bottom row indicates how quarks interact with each other via the strong force by exchanging a gluon. (Color version of the top right on Page C-2.)

Before we continue, it is worth saying a bit more about the electric charge of the particles we have mentioned. Prior to the 1970s, scientists thought all charged particles had the same magnitude of charge since that is what had been observed in nature. They were either negative or positive versions of the same amount of charge. While most people are comfortable saying an electron has a negative charge, or a charge of -1, scientists would say an electron has the same charge as the proton but times negative one. Since most people are not scientists, particles are more commonly described as having a charge of plus or minus one, or sometimes just positive or negative.

However, as scientists learned more about quarks, they discovered that their assumption about unvarying charge magnitude in particles was not true. Using more sophisticated experiments of the type shown in Figure 7.1, they were able to determine that quarks can have smaller amounts of charge.

We now know that a nucleus is a combination of protons and neutrons that are, in turn, made up of quarks (Chapter 3). While it isn't strictly true, we can simplify and say that a proton is basically comprised of two different types of quarks, known as "up" quarks (each with 2/3 of the value of the charge of a proton) and "down" quarks (with a charge of -1/3 of the proton). To see how this works out, the charges of two up-quarks and one down-quark combine for a total charge of +1; note we stopped saying "times the charge of the proton" for simplicity's sake. Similarly, a neutron has two down-quarks and one up-quark, so the charges cancel out and we get a zero, or neutral, charge.

This leaves a simple question: Why don't the two up-quarks in the proton, which are really close to each other and have the same sign, repel each other and break the proton apart? The answer is that the attraction they feel for each other due to the strong force is more powerful than the repulsion of electromagnetism.

This force is also well-described using quantum mechanics. In the same way that a photon is the "mediator" between two electrically charged particles, the **gluon** is the particle that carries information from one quark to another. Gluons "glue" quarks together. [1] Put another way, quarks "talk to each other" via gluons (see the bottom part of Figure 8.1). [2] Again, note that the amount of electrical charge before and after the interaction stays the same. The presence of gluons in both protons and neutrons is but one reason our simplified up-and-down quark description of these particles' makeup isn't strictly true.

---

[1] Gluons, like photons, are also massless and always move at the speed of light.

[2] We note for now that there is another force, the **weak force**, which is mediated by the W and Z particles (see Table 3.1). We will disregard these two for now, but will meet them again in Chapter 19.

The gluons that keep quarks together in a proton also keep protons in the nuclei from being forced apart and thus breaking up the nucleus. Specifically, the quarks in the proton not only attract each other, but also attract the quarks in any nearby proton or neutron. In other words, the strong force basically "leaks out" of protons and neutrons, thereby keeping them all together in the nucleus.

Thanks to the strong force, there are many ways in which protons and neutrons can combine. To help scientists keep things straight, we give each combination a different name, which comes from the number of protons in the nucleus.

For example, an atom with only one proton is called hydrogen (H). An atom with two protons is called helium (He). A helium nucleus has a charge of +2, compared to hydrogen's charge of +1. This means helium can accept two electrons in orbit before becoming electrically neutral. When it becomes neutral, it will stop pulling in more electrons or other atoms, and stop repelling more protons. Since hydrogen has only one proton, it can accept only one electron in orbit before it becomes neutral.

The way an atom interacts with the rest of the world, then, critically depends on the number of protons in its nucleus. To illustrate the makeup of these atoms, scientists use the **periodic table of elements,** which you may remember from your middle school or high school chemistry books.

In addition to protons, atoms heavier than the lightest element, hydrogen, typically contain neutrons in the nucleus. Since both the number of protons and the number of neutrons in the nucleus alter the behavior of the atom, we label atoms with different numbers of neutrons as **isotopes.**

For example, a hydrogen atom that has a single neutron is known as the isotope **deuterium,** or ^{2}H (the superscript labels the combined number of protons and neutrons). Similarly, adding one or two neutrons to helium gives us the isotopes ^{3}He and ^{4}He. We can add as many protons and neutrons as we like, in principle,[3] and get atoms such as carbon (whose isotope ^{14}C is used in radio dating), oxygen, and nitrogen (which are both abundant in the air we breathe).

As predicted by quantum mechanics—and observed in experiments—atoms with different numbers of electrons, protons, or neutrons have various available energy levels for their electrons. When electrons move among these energy levels, they produce and absorb different colors of light. This is especially useful to detectives because it means that each type of atom absorbs only certain colors (wavelengths) of light, which produce different spectral lines, as shown in Figure 8.2.

---

[3] Typically, the number of neutrons is the same as the number of protons. Also, when there are more than ninety protons in a nucleus, most nuclei are not stable and cannot live very long.

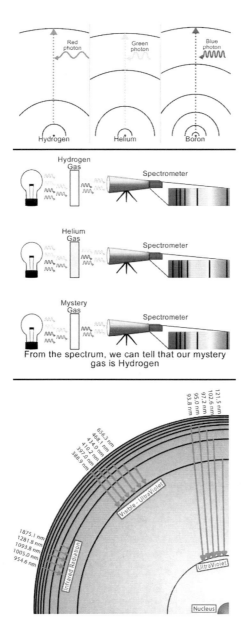

FIGURE 8.2 Different types of atoms have different energy levels. As the white light shines on the atoms, the electrons move between these levels and absorb only special colors. Since the energy levels are so distinct, this allows us to "fingerprint" the different types of atoms. The large number of the different photon wavelengths that come from hydrogen gives us confidence that we understand how atoms work. There are lots of lines to look for; we have observed them all at the very places quantum mechanics predicts they should be. (Color version on page C-6.)

As our expert lab technician knows, we can shine white light at hydrogen atoms here on Earth and observe in our spectrometer that a particular set of colors is absorbed (and there are many of them, as shown on the right side of Figure 8.2). Helium has a completely different set of spectral lines. A third set will come out of oxygen, and so on.

We can also go the other way. If we use a spectrometer to observe the photons passing through a mystery gas and note that only the colors typically absorbed by hydrogen are missing from the resulting spectrum, we can safely conclude that the gas is hydrogen. Similarly, if we note that only the colors absorbed by helium are absent, then we have confidence that the gas is helium. This makes our atomic fingerprinting even more powerful.

So by looking at the light from stars, we can not only determine that they are comprised of atoms, but even can tell which types of atoms they are made of and how many there are of each. In addition, the fact that there are no spectral lines we don't recognize gives us further confidence that we really understand what is going on.

8.2 THE STUFF IN THE STARS: WHY THEY SHINE, HOW THEY MOVE, AND HOW WE CAN TELL

We now know that stars are giant balls of (mostly) hydrogen atoms. What is the evidence for this? Does this description explain why we see lots of light coming from stars?

To answer these questions, we must determine the following: What would happen inside a giant ball of hydrogen? And if we assume a star is a giant ball of hydrogen, does it predict the kind of light and spectral lines we see in the data? Since we know that hydrogen atoms would interact and collide with each other, we look at simple collisions among hydrogen atoms here on Earth to get an idea of what happens when they interact in space. These are shown in Figures 8.3, 8.4, and 8.5, where we ignore the electrons for simplicity.

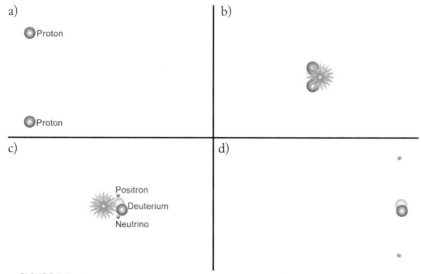

FIGURE 8.3 A series of action shots (see also Figures 8.4 and 8.5) that show nuclei interacting in a star and giving off light. On this series, two protons experience a nuclear interaction (symbolized by the energy burst). As they fuse together, one of the protons turns into a neutron, a positron, and a neutrino (a very light neutral particle mentioned in Chapter 3, which we will come back to in Chapter 19). The proton and neutron pair is known as deuterium, or ^{2}H. After the interaction, the three particles move off into space. Because this is a nuclear interaction, all three come out with a lot of energy. The positron will eventually meet an electron in the star and produce a pair of high energy photons.

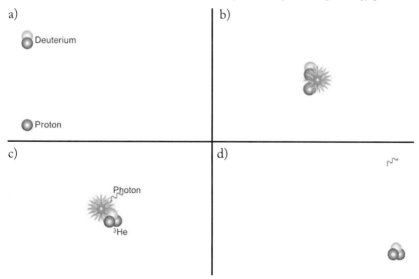

FIGURE 8.4 A set of action shots similar to Figure 8.3, but here we see a deuterium nucleus interacting with another proton to produce ^{3}He and a photon.

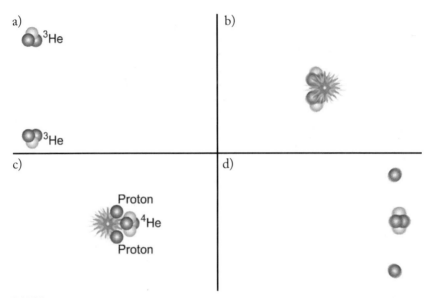

a)

³He

³He

b)

c)

Proton

⁴He

Proton

d)

FIGURE 8.5 A set of action shots similar to Figures 8.3 and 8.4, but here we see a pair of ³He nuclei, which combine to form ⁴He and two hydrogen nuclei. These two hydrogen nuclei are used to start the process in Figure 8.3 all over again.

If there is a large number of hydrogen atoms gathered together, as there is in a star, many of the atoms will fall toward the center like a rock falling to the surface of the Earth. When this happens, an atom can become very energetic.

When two very energetic protons get near each other, they can collide and turn into deuterium, or ²H. As shown in the figure, one of the protons turns into a neutron, a positron, and a neutrino—a very light, neutral particle.[4] This interaction between two nuclei is referred to as a **nuclear interaction** or a **nuclear reaction**. In general, we call the combining of any two nuclei **fusion**. This phenomenon gives each of the three particles a lot of energy. It is natural to ask where this energy comes from. We note that the mass of two protons is larger than the mass of the deuterium, the positron, and neutrino; this is an example of mass being converted into energy using $E=mc^2$.

We now have deuterium in the star, and when a proton and deuterium bang into each other, they, too, have a nuclear reaction. This produces ³He and a photon. We can also have two ³He collide to produce ⁴He. There are other types of interactions as well, but essentially, the enormous number of nuclear reactions in a star produces the light (photons) we see. This is why stars shine.

---

[4] We mentioned neutrinos in Chapter 3, we will come back to them in Chapter 19. It is also important to note that the neutrino is literally created during this process. It was not hiding inside the proton. This is a quantum mechanics thing.

Nuclear reactions in a star produce light at different energies. The photons bounce around inside the star, changing their energy/color before they come to our eyes. While it is more complicated than this, one of the reasons our Sun looks like a mostly-yellow light bulb is because there are a plethora of different color photons reaching our eyes. We will discuss this more in Chapter 16.

While light produced through the interactions of nuclei can explain why stars shine, we look to atomic fingerprinting for our next piece of evidence.

Think about the light produced in the inner part of the star (called the **core**). If we consider what happens to light as it passes through the atoms in the outer part of the star on its way to our eyes, we quickly realize that some of the photons will interact with these atoms if they have the right energy/wavelength. When this happens, the photons are absorbed, and the atom transitions into a higher-energy state. Since these photons are absorbed, they will never hit our eyes.[5]

If we aim our telescopes—with spectrometers attached—at a star, the colors that were absorbed should appear as dark lines in the spectrum seen in our spectrometer. And that is precisely what we see. The fact that we observe all the expected spectral lines—and no unexpected ones—helps confirm that a star is a giant ball made up of atoms, as shown in Figure 8.6.

Taking into account all the complicated properties of stars, astronomers have learned that our Sun is overwhelmingly made up of hydrogen and helium. While these aren't the most common elements here on Earth (that honor goes to oxygen and iron), hydrogen and helium turn out to be the most abundant elements in the universe, and the most common components of stars everywhere—although the amount of each type of atom can vary considerably from star to star. Then again, atoms are only a small fraction of the mass in galaxies.

For now, what is important to realize is that our Sun is made of the same material as the other stars. This is quite remarkable. Even when we look at stars billions of light years away—which means they sent their light billions of years ago—we observe that they have the same spectral lines we see from nearby stars. This provides excellent evidence that much of the "stuff" out there is made up of the same "stuff" we see on Earth. It gives us confidence that physics is likely to be the same everywhere—and has been for billions of years.

One final thing about stars before we finish our chapter: by looking at the spectral lines from a star, we can learn about the speed of a star using the Doppler effect

---

[5] It is true that the electron in the atom will eventually move down to a lower-energy level and emit its signature photon. However, since the light can come out in any direction, it is unlikely that the photon will come directly at our eyes. We will therefore never see it.

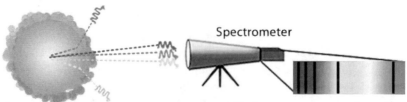

From the spectrum, we can tell the star is made of Hydrogen gas

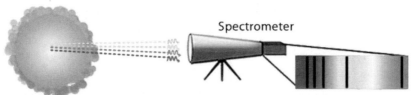

From the spectrum, we can tell the star is made of Hydrogen gas and moving away from us

FIGURE 8.6 A simplified view of looking at light from a star. By looking at the light from stars using a spectrometer, we can tell what they are made of and how they move. Note that the spectral lines in the bottom figure are shifted to the right (red-shifted) relative to the top figure. (Color versions on page C-6.)

(Chapter 5). To understand how spectral lines can be Doppler-shifted, consider a photon as it hits an atom. If the photon has an energy that corresponds to the change of two energy levels, it will be absorbed. But if the atom is moving toward the light, then only photons with a smaller amount of energy (longer wavelength) will be absorbed. From the atom's perspective, the light's wavelength has been Doppler-shifted to a different color/higher energy, which is why it is absorbed, as shown in Figure 8.6.

So if a star is moving, then all the atoms in the star will be moving; thus, we will see the spectral lines in different places. Stars moving toward us will have blue-shifted spectral lines. Stars with red-shifted spectral lines are moving away from us. We can determine much about the motion of stars and galaxies from looking at spectral lines undergoing Doppler shifts. Indeed, we can measure the speed of the star quite accurately with modern equipment.

We now leave the stars and what they can tell us, and move on to study temperature and thermal equilibrium. With this knowledge in hand, we will be able to understand more about the history of our universe, the role of the earliest times after the beginning of the universe, and the story of what transpired between then and now.

Temperature and Thermal Equilibrium

As we saw in Chapter 2, the majority of the space between stars is not very dense. Indeed, it is filled only with an occasional particle like a hydrogen atom, a photon, or dark matter.

To visualize this almost entirely empty space, think of a balloon filled with red smoke that has just popped inside a room. You can see the red smoke clearly for a little while, but eventually, it spreads out and fills the room. The smoke molecules may be harder to see, but they are still there.

To understand outer space and the hard-to-see atoms and photons in it, we will study much smaller collections of these particles here on Earth. To describe these particles, we will start by using important words like **temperature** and **thermal equilibrium** in situations we already experience.

As a simple first example, consider a bunch of atoms moving around in a room. As they travel, they interact with each other. If I have a variety of atom types in a room and if they can mix—like the red smoke mixing with the air—they eventually will do so.

If the atoms start with different temperatures on opposite sides of the room (hot on one side and cold on the other—more about what hot and cold really mean will be explained in greater detail soon) the mixing process will cause the room to have the same temperature throughout (if someone has not left the window open!). We call this resulting state **thermal equilibrium.** We will describe it—and temperature—in more detail later in this chapter.

As we will see, this phenomenon is not just true for atoms in a room; any type of particle can interact and eventually come into thermal equilibrium. As detectives, this is a critical concept for us to understand. In essence, we can predict how any given set of particles will change over time, and what this change will ultimately look like after the particles have reached thermal equilibrium. We can even do so for the particles in the entire universe.

For now, though, let's make the concept of thermal equilibrium a bit less overwhelming by confining our set of particles to a room. Since we are used to talking about the temperature of a room—and this is a central concept in what a room in thermal equilibrium will look like—we start there.

9.1 What is temperature?

TV weather reporters talk about the temperature. Americans tend to describe the temperature in Fahrenheit. People everywhere else use Celsius—except scientists, who use Kelvin. You can think of room temperature as roughly 70 degrees Fahrenheit, which is about 30 degrees Celsius. Since Kelvin is larger than Celsius by 273, room temperature is approximately 300 Kelvin (we don't use the word degree to describe Kelvin, and we abbreviate Kelvin with a K).

But what do we really mean by temperature? Typically, we think of it in terms of "feeling hot" or "feeling cold." When we feel cold, the thermometer displays a lower number. On sunny summer days, we feel hot and the thermometer displays a higher number.

Why do we feel hot or cold? What does a thermometer measure?

To answer these questions, think about what happens during the daytime. Photons hit your skin and are absorbed by your body. Each of these photons can deposit its energy into the atoms in your body. This makes your body heat up and you feel this heat.

Let's take this idea a bit further. An ultraviolet photon has more energy to deposit than an infrared one. Thus, if your body absorbs all the light that shines on you, the ultraviolet light will make you feel hotter than the infrared. When it comes to temperature, what matters is the amount of energy your body absorbs (see Figure 9.1); the more energetic the photons (which have shorter wavelengths), the higher the temperature.

Note that your skin is not a good thermometer. You feel hotter in direct sunlight than in the shade because more photons are hitting you. This is misleading; the temperature is typically no lower in shade than it is in direct sunlight.

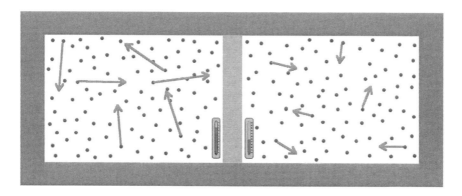

FIGURE 9.1 Two atom-filled rooms illustrate the difference between high and low temperatures. Each room has only one type of atom (helium, for example). The length of each arrow (only on some of the atoms) indicates how fast that atom is moving, and thus its energy. In the room on the left, each atom has a fast speed, which means it has a high energy. The atoms in the room on the right, however, have a slower speed, indicating a smaller energy. A person inside the room on the left would therefore feel a high temperature, while the person in the other room would feel a low temperature.

We can talk about temperature in terms of a bunch of photons or a bunch of atoms. Since atoms, like photons, have energy and can hit you, a group of atoms has a temperature. So even if we are in a perfectly dark room with zero photons, we can still feel warm. This is because our body is coming into contact with atoms, which are depositing energy into our skin that heats us and makes us feel warm. As with photons, the more energetic the atoms, the higher the temperature.

If we have two rooms, each filled with atoms of different speeds, then we can determine which of the two rooms has a higher temperature. We can talk about the temperature of a star in the same way. If a star has a high temperature, we know that the atoms in it are very energetic and most of the photons coming from it will possess the same energy. The temperature of the surface of the Sun, for instance, is about 5,800 K; we primarily see the yellow light it emits.

9.2 THERMAL EQUILIBRIUM

Atoms can have different temperatures in different parts of the same room. For example, if you are in a hot room and someone turns on the air conditioner for twenty minutes, then some of the atoms in the room will have lower energy than others. Assuming there are no windows letting in light or outside air, everything eventually evens out and the whole room comes to a (hopefully) reasonable temperature. If there is a fan in the room, the cool air mixes more quickly with the hot air. In principle, once things are fully mixed, they would stay this way forever.[1]

What is happening at the atomic level during the mixing process? It turns out that both simple and complicated things can happen. For example, atoms can simply bang off each other like billiard balls. However, because of quantum mechanics, other things could also happen. The atoms, for instance, could come apart or release photons (like we discussed in Chapters 7 and 8). Even more interesting, if there are both particles and anti-particles in the room (like we discussed in Chapter 3), we could have annihilations where the pairs turn into light. We will consider this type of interaction soon enough, but for now, we start with atoms colliding in the room in which you are currently sitting.

At room temperature, individual atoms bang into each other like billiard balls. Consider a pool table where a moving cue ball strikes a stationary pool ball. After the collision, the stationary ball begins to move and the cue ball travels away at a slower speed. This is true even if the second ball is not stationary, but rather moving. In general, a high-speed ball that collides with a slower-moving ball will become slower, while the slower-moving ball will actually gain speed.

The same is true for simple collisions between atoms. When a high-energy atom strikes a low-energy atom (Figure 9.2), it can "transfer" some of its energy. As a result, the low-energy atom speeds up and becomes a medium-energy atom. The opposite is also true: the higher-energy atom gives up some of its energy and therefore slows down, also becoming a medium-energy atom. After the collision, both atoms move with speeds that are, on average, closer together.

How does this explain why the room eventually becomes the same temperature throughout? Consider an extreme example.

At the beginning of a game of pool, right after we shoot the cue ball at the full rack of balls (see Figure 9.3), the cue ball (our high-energy atom) starts with high energy, while the others start with low—or zero—energy. After they collide, the cue ball has a lower energy than before, while the regular balls now have some energy of motion. In the collision, the energy was transferred from one ball to the others. Now all the balls on the pool table move around and collide.

---

[1] Unfortunately, in everyday life, the temperature outside the room changes and, because no one has perfect insulation, the room temperature eventually changes with it. In the rest of this chapter, we are going to be talking about atoms and other particles inside a room (or other containers) where the "outside" doesn't influence the particles inside.

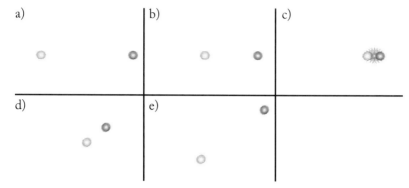

FIGURE 9.2 A series of action shots showing the collision between a high-energy atom and a low-energy atom. The high-energy atom on the left has a fast speed and is moving toward the low-energy, and thus slow-speed, atom on the right. After the collision, the atom on the left has a lower energy (and speed) than it used to, and the atom on the right has a higher energy (and speed) than it used to. Overall, the speeds are now closer to being the same.

If we had a perfect pool table (one without friction), the balls would move around forever, with the high-energy balls bouncing into the low-energy ones until they all became medium-energy balls. It is just a matter of time, then, before a large number of collisions occur and all the balls have roughly the same energy.[2] At that point, things will stop changing.

The same is true of temperature. When a group of atoms shares an identical level of energy, the temperature will come to be virtually the same everywhere. This situation is called **thermal equilibrium**.

By way of contrast, let us look at a similar scenario that is *not* in thermal equilibrium.

The bottom set of rows in Figure 9.3 shows the same pool table and pool balls, but now we have a bad shot: the cue ball will *never* hit the rack. Instead, the cue will always stay high-energy and the other balls will never get any energy, so this system is *far* from equilibrium.

How can I tell if I am in a room in thermal equilibrium? If I am sitting in a room in thermal equilibrium, there should be an equal number of atoms with roughly the same energy coming at me from all directions.

---

[2] We note that just like the balls on a pool table, some atoms move faster than average and some move slower than average. We are able to predict both the average speed, as well as how many (on average) possess each speed. This is known as the **Boltzmann distribution.** It can be used to measure the temperature of the balls on the table—or atoms colliding in space. For more detail, see the books in the Suggested Reading.

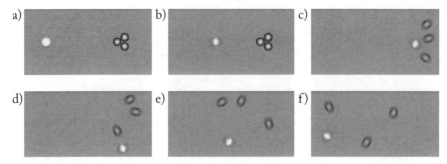

FIGURE 9.3 A set of action shots showing balls moving on a perfect pool table where friction does not slow them down. In the top two rows, we start with a scenario where the balls are not in thermal equilibrium. Starting in "b," the cue ball has been shot and moves toward the rack. In "c," the cue ball has hit the regular balls and they begin to bounce around. Eventually, by the last figure, they all end up moving at roughly the same speed and in random directions; we say the balls are effectively in thermal equilibrium. The three rows in the bottom part of the figure show a very different scenario. In this case, the cue ball never hits the rack and the balls on the table never reach thermal equilibrium.

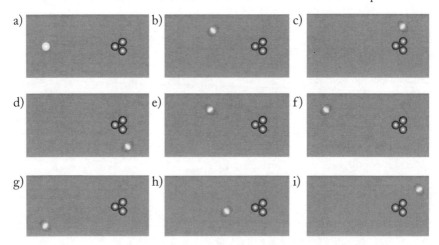

Think of a diver in the middle of a school of fish (see Figure 9.4). From the diver's perspective, since the fish are swimming at about the same speed in all directions, the "fish temperature" is the same in every direction. What if the diver is swimming to the right? Then there would be an equal number of fish in all directions, but the speed of the fish would vary in different directions; they would be moving toward the diver faster in the direction she is swimming, and slower in the opposite direction. Thus, the "fish temperature" would vary in different directions, even though the fish are in "thermal equilibrium."

BIG BANG, BLACK HOLES, NO MATH

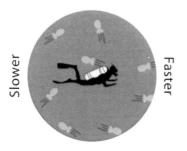

FIGURE 9.4 A diver in a school of fish. The fish can be thought of as being in thermal equilibrium if they are all moving in random directions and all have roughly the same speed. The speed of the fish from the diver's perspective is different in the two scenarios, but she can still tell that the fish are in "thermal equilibrium."

Like the diver in the school of fish, you can tell you are in a room that is in thermal equilibrium if the temperature is the same in all directions, or if the temperature looks as if you (or the room!) are moving in a single direction. In a room that is perfectly insulated, it is only a matter of time before the photons and atoms in the room reach thermal equilibrium.

Similarly, if atoms or photons are in thermal equilibrium, they remain so forever. Of course, there are exceptions we should note. Atoms held in place inside a rock, for example, are not able to interact and come into thermal equilibrium. Likewise, if a room is not perfectly insulated and cold air comes inside through leaky windows, the atoms may never come into thermal equilibrium, or may leave thermal equilibrium when the temperature outside changes.

9.3 THERMAL EQUILIBRIUM WITH OTHER TYPES OF PARTICLES

We have talked about the simplest case of atoms interacting and coming into thermal equilibrium, but many other types of particles interacting in different ways can also reach thermal equilibrium. For example, the room you are sitting in now is filled with air that contains both oxygen and nitrogen atoms. Let us consider how it might have come to be that way.

In a simplified example shown in Figure 9.5, high-mass and low-mass atoms are initially independent and have different temperatures. Since temperatures on each side vary, it is clear that the atoms are not in thermal equilibrium.

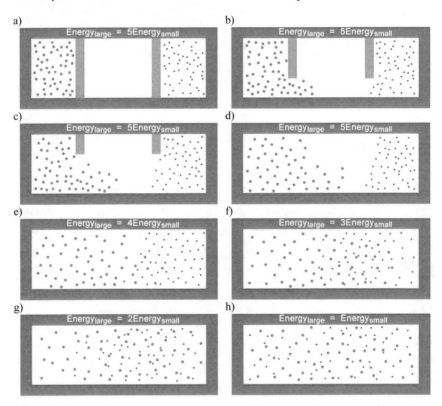

FIGURE 9.5 A set of action shots showing two sets of atoms mixing. While the same number of atoms are present on both sides, the atoms on the left initially have, on average, five times more energy than the atoms on the right. After the barriers between the two are removed, the atoms bounce off each other. Ultimately, the atoms mix and their energies change until they are all roughly the same. At that time, the system has the same temperature everywhere and has come into thermal equilibrium.

When the atoms start to mix, they interact. It is just a matter of time until a sufficient number of collisions take place, everything is mixed, and the atoms share the same temperature. At this point, the room has reached thermal equilibrium.

It is important to note that even if all atoms in a room share the same temperature, this does not necessarily mean they are in thermal equilibrium. For example, consider a room with two different kinds of atoms. Each of the atoms has the same temperature, but one of the atom types is on one side of the room, and all of the other atoms are on the opposite side. The room will not be in thermal equilibrium until all the atoms completely mix and intermingle.

Since atoms are not the only things that can be in thermal equilibrium—and simple billiard-ball collisions are not the only types of interactions between particles—it is time to discuss other types of collisions. Keep in mind that any kind of collision can occur as long as it follows the laws of physics (for example, the amount of charge before the collision is the same as the amount of charge after the collision). This is important since collisions can actually change the types of particles in a room.

In Chapter 7, we looked at some special ways that electrons and photons interact. But quantum mechanics allows other types of collisions as well, such as those shown in Figure 9.6. In the figure, two very high-energy photons collide and turn

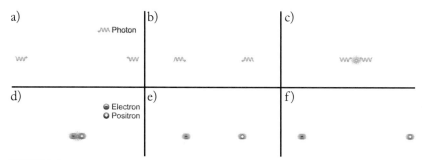

FIGURE 9.6 A sequence of action shots where two high-energy photons hit each other and turn into an electron and a positron (an anti-electron). In the bottom sequence, we see the inverse, where an electron and a positron hit each other and turn into a pair of photons. Note that these types of interactions only occur when the energies are very high, as they were right after the big bang or today's particle accelerators.

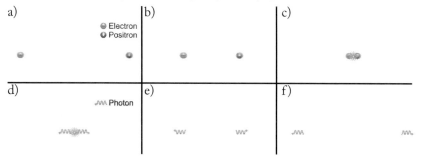

into an electron and an anti-electron (also called a "positron," as we saw in Chapter 3). Note that this does not usually happen in the room in which you are sitting because the typical photons you see with your eyes have an energy ten thousand times too small for that to occur.

A second collision example is that of an electron and an anti-electron colliding and turning into a pair of photons.[3] Just as high-energy and low-energy atoms can interact to produce medium-energy atoms, so electrons and positrons can interact to create high-energy photons, and vice-versa.

When trying to figure out what will happen to a room as it moves into thermal equilibrium, it is important to remember that particles in the room can interact and change from one type to another. Since high-energy photons can create electrons and positrons, and electrons and positrons can create photons, they can all mix. It is not just that the energy of all the particles will eventually be similar, but that the number of each *type* of particle will come into equilibrium.

Let us illustrate this phenomenon with an extreme example.

Consider a room filled solely with electrons and positrons that share the same temperature. The electrons and positrons in the room continually collide, producing photons. Since the situation in the room continues to change, we know the room is not in thermal equilibrium.

Oddly enough, we will never run out of electrons and positrons. At some point, there will be enough photons so that they collide with each other and complete the circle by producing more electrons and positrons (see Figure 9.7). But eventually, the number of electrons, positrons, and photons will stabilize, and we will reach thermal equilibrium.

Similarly, if we had started with only very high-energy photons with the same temperature as in the above electron-and-positron-only case, electrons and positrons would soon be produced. Again, the whole room will eventually come into thermal equilibrium. Both cases end up with the same temperature throughout; we will even arrive at the same number of electrons, positrons, and photons. What ultimately determines the final number of each type of particle and its energy is the temperature—not just the initial particle makeup of the room.

---

[3] This type of interaction is probably where science fiction TV shows get their ideas for anti-matter guns. If I shoot anti-matter—such as an anti-electron—at a target and it hits an electron in an atom, then the electron/anti-electron pair could turn into a big burst of light (two photons). We note that the amount of charge before the collision is zero (the sum of the positive charge and the negative charge) and there is no net charge after the collision since both photons are neutral.

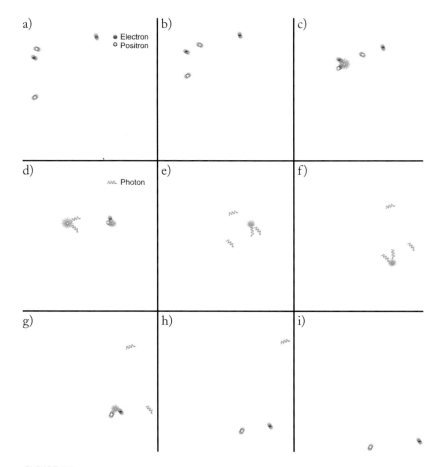

FIGURE 9.7 A series of action shots showing photons, electrons, and positrons in thermal equilibrium with a very high temperature. Starting in the top left figure, two electrons and two positrons approach each other. Eventually, both pairs collide and create a pair of photons. One of these pairs of photons then collides and creates another electron and positron pair.

In fact, if a scientist knows the particle types and their energies, she can predict the average number of each type and the average energy (or temperature) after the room has come into thermal equilibrium. In some cases, she can even tell how long it will take before the room reaches equilibrium.

Ultimately, it is possible for most rooms full of particles to eventually come into thermal equilibrium. If we know the temperature of a room, we know what the room will look like after it comes into thermal equilibrium. If we are in a room, we can tell if it is (or was) in thermal equilibrium.

9.4 What We Can and Cannot Learn from Systems in Thermal Equilibrium

There is good news and bad news about groups of particles that come into thermal equilibrium. The good news is that when you close the door to your room and turn on the air conditioner, no matter the starting temperature of the atoms, it is just a matter of time before you get a cool room. The bad news is that if many different initial temperature conditions all lead to the same thermal equilibrium temperature, we cannot, as detectives, work backward and determine how things began. In other words, we cannot learn for certain what happened *before* the room came into equilibrium.

So if we walk into a room that is in thermal equilibrium, we have no way of knowing whether the room started out cool, or if a fan caused hot and cool air on opposite sides of the room to mix. To figure it out would require a clue or some other piece of evidence.

Let's take another look at our pool table, this time with the balls in equilibrium. We cannot tell if a single cue ball hit the full rack of balls, or whether the rest of the balls started in some other configuration (see Figure 9.8). If all we can see is how things presently look (picture "e"), we cannot tell how they started (picture "a").

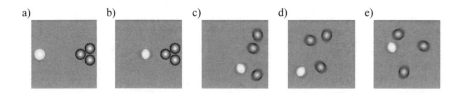

FIGURE 9.8 Two sets of action shots showing scenarios that come into thermal equilibrium. In the top sequence, the balls start on the table in the usual way, while in the bottom sequence, they start in an unusual way. In both instances, however, the balls come into thermal equilibrium. If we just see the fifth picture in either sequence, it is virtually impossible to tell how things started. If you take into account the uncertainty from quantum mechanics, it is completely impossible to know.

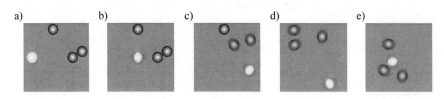

In the next few chapters, we will see that the universe acts like a giant room of various particle and atom types that at some point was in thermal equilibrium. By understanding rooms that are, or were, in thermal equilibrium, we can tell a lot about the universe and how it changed over time.

The problem, however, is that once the universe came into thermal equilibrium, we lost some of our ability to learn what happened beforehand. Much like the pool ball example, we cannot yet say anything with confidence about the beginning of the universe—what we are calling the "big bang." While we would like to know what happened at the beginning, or before the beginning (or even if there *was* a "before the beginning"), all we can describe confidently is what happened *after* the big bang and after the universe came into thermal equilibrium.

Now that we have finished going over the physics we will need (Unit 2) to solve these mysteries, we will use what we have learned to understand the evidence for the big bang (Unit 3).

The Evidence for the Big Bang

In this unit, we describe the evidence for the big bang and how the universe changed from its early beginnings into the one we have today. We will concentrate on the experimental evidence and how scientists interpret it. Each of the following three chapters covers a separate set of topics, but most of the ideas are interrelated. They are:

Chapter 10: The Exploding Universe
Chapter 11: Expanding Space-Time
Chapter 12: Photons and Hydrogen in the Universe

How we interpret the scientific clues in these chapters may seem complex at first, but I hope to de-mystify the whole process. The three primary pieces of evidence to explain are:

1. **The galaxies in the universe**: We observe about a hundred billion galaxies in the universe and note that they seem to be distributed evenly throughout. More importantly, we see that all of the distant galaxies are moving away from us. The farther away the galaxy, the faster it moves away.
2. **The photons in the universe:** Low-energy photons (microwaves) are uniformly distributed in all directions with a temperature of about 2.7 degrees above absolute zero (Kelvin). It is as if they are, or were, in thermal equilibrium.
3. **The atoms in the universe:** About 91 percent of the atoms in the universe are hydrogen, while most of the rest are helium. There are precious few atoms of any other type.

In Chapter 10, we discuss light from galaxies and how it leads us to the theory that we live in a universe that began with a big bang and has been expanding ever since. As the data can be misinterpreted to point away from a big-bang interpretation, we discuss how general relativity and expanding space-time come to the rescue in Chapter 11.

In Chapter 12, we talk about the photons in the universe, known as the cosmic background radiation, and the fact that most of the atoms in the universe are hydrogen and helium. While many consider these last two pieces of evidence to be the most compelling, I find it hard to explain why they are evidence at all to someone until they know enough of the story of the universe's evolution from the big bang to the present. Thus, we will explain them after a quick discussion of the expanding universe.

The Exploding Universe   CHAPTER 10

Before the 1920s, there was a small amount of data that suggested that some of the light from "things out there" might be from galaxies beyond our own. However, the evidence was not convincing. Many scientists, including Einstein, thought our galaxy was the entire universe.

Astronomer Edwin Hubble would soon prove otherwise.

Armed with the most powerful telescope of its day, Hubble firmly established that there is a giant universe beyond the stars of our galaxy, the Milky Way. We now know that just as our Sun is only one of many stars, the Milky Way is just one of many galaxies.

In this chapter, we will talk about the data that comes from looking at these distant galaxies. We will then discuss how this provides evidence that we live in a fascinating universe that appears to have exploded into existence and has been expanding ever since.

10.1 WHAT CAN WE LEARN FROM LOOKING AT GALAXIES?

After establishing that there are many distant galaxies, Hubble studied them to answer such questions as: How far away are they? In what direction are they moving? How fast are they moving?

Let us consider these questions one at a time.

There are several complex techniques utilized to measure the distance of a galaxy. The one employed by Hubble was to study a special type of star in a galaxy, known as a **Cepheid variable**, because it is both very bright and very well understood.[1]

---

[1] The amount of light that comes from a Cepheid variable goes up and down over time; it varies, thus the word "variable" in its name.

We can think of each Cepheid variable star as a light bulb; some are 25 watt bulbs, some are 100 watt bulbs, etc. and we can determine the wattage of each. This helps because if we have a bulb with a known wattage, then we can measure how "bright" it appears (by counting how many photons we see coming from it) when it is ten meters away, fifty meters away or a hundred meters away. Conversely, if it is some unknown distance away, then by counting the number of photons we see, we can determine how far away it is. The same is true for a Cepheid; since we know how many photons each is emitting and we can count how many we see, we can figure out how far away it is (see Figure 10.1).

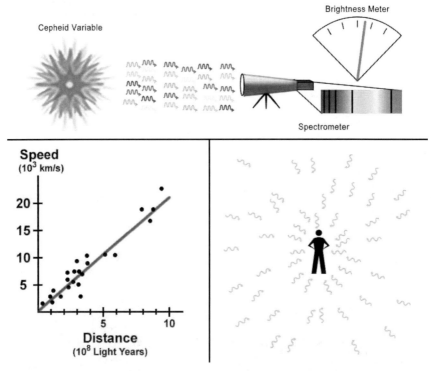

FIGURE 10.1 An artist's conception of one of the techniques to measure both distances to galaxies and their speeds and directions. Here, our simple brightness meter is counting the number of photons observed in the spectrometer. By looking at many stars in a galaxy, we can study the amount of light from them—as well as their Doppler shifts—to see whether the galaxy is moving toward us (blue-shifted) or away from us (red-shifted). In the bottom two figures, we see that the more distant the galaxy, the faster it is moving away from us. To use this plot to determine how far away a galaxy is, pick a dot on the plot and go straight down to the horizontal axis. Reading off the number tells you its distance from us. Similarly, if you pick the same dot and go to the left until you hit the vertical axis, you can find its speed away from us. We also have drawn a line that describes how the dots fall in the plot. It shows the relationship between the speeds and the distances, known as Hubble's Law. In the bottom right, we see that the data are basically the same in all directions. (Color version on page C-7.)

Taking it a step further, if we find one or more of these special stars in a galaxy, we can figure out the distance to that particular galaxy. One of the most well-known nearby galaxies is Andromeda. Using Cepheid stars in Andromeda, astronomers determined it to be about two million light-years away.

Remember that the Milky Way is only a hundred thousand light-years across (see Figures 2.3 and 2.4). To put yourself in the shoes of Hubble when he realized Andromeda was so far away, it is like having grown up in Houston, thinking that Texas is the entire universe, and then realizing that there are cities in New York, California, and deep into Mexico. The farthest galaxies we now know of are more than ten billion light-years away—farther than cities in India in our analogy.

How do we tell in what direction and at what speed the galaxies are moving? We can observe the light the galaxies emit, and then use the Doppler shifts of the spectral lines to learn about the motion toward or away from us (again, see Figure 10.1). Ignoring side-to-side motion of galaxies for now, we expect that if a galaxy is not moving, there is no shift in the wavelength of the light from it. But if the light is shifted toward the blue end of the spectrum (a blue-shift), then we can tell that the galaxy is moving toward us. If the light is red-shifted, the galaxy is moving away from us.

When we look at galaxies throughout the universe, we find that:

1. There are about a hundred billion galaxies in the universe and they appear roughly evenly spread in all directions.
2. All distant galaxies appear to be moving away from us (red-shifted).
3. More distant galaxies appear to be moving away from us at a faster rate.

It is worth saying more about this data.

If we look to the right and see a galaxy that is ten billion light-years away, it will be moving to the right (away from us) with a high speed. If we look to the left and find another galaxy that is ten billion light-years away, it will be moving with roughly the same speed, but this time it will be moving to the left (again, away from us). The same will be true in any direction we look.

These facts are shown in the plot in Figure 10.1. To use this plot to figure out the distance of a galaxy, pick a dot and go straight down to the horizontal axis. Reading off the number tells you the galaxy's distance from us. Similarly, if you pick the same dot and go to the left until you hit the vertical axis, you can find the galaxy's speed away from us. We also have drawn a line that roughly describes how the dots fall in the plot by considering the relationship between the speeds and the distances. This relationship is known as **Hubble's law.** We will come back to why all the dots are not on the line in the next chapter.

Note the large size of the values on both axes. Galaxies are apparently traveling away from us at thousands of kilometers per second. To put these numbers in perspective, entire galaxies are moving at speeds so fast that they could travel from New York to California in about one second!

You should also pay attention to the incredible distances illustrated by this plot; the farthest galaxies are more than ten billion light-years away. To put this in context, the most distant stars in our galaxy are about a hundred thousand light-years away from Earth (and the star closest to our Sun is about four light-years away). Since this is 10^5 light-years, or 0.001×10^8 light-years, it would be found only in the lower-left-most portion of the figure.

As the stars in our galaxy are not moving away from us a great deal, it's easy to understand why scientists before Hubble's discoveries thought the universe did not change much. Ultimately, the remarkable results about distant galaxies showed that not only is the universe much bigger than a single galaxy, but also that everything in it appears to be rushing apart! Is something *pushing* it?

Before we move on to an explanation of this data, we also note that new data from the 1990s indicates that not only are the galaxies quickly moving away from us, but also that they are accelerating as they do so. We will come back to this data when we discuss dark energy in Chapter 18. For now, we will focus on the observed facts that the galaxies are moving away from us, and that the speeds at which they move are faster the farther they are from Earth.

10.2 A Simple-but-Wrong Model of an Exploding Universe

A proper understanding of the motion of the galaxies was first described in the late 1920s by Georges Lemaître, a scientist who was also a diocesan priest. Lemaître was trying to understand the galaxy data and wanted to use general relativity to describe a universe that changes over time. He hypothesized that galaxies do not rush apart because something "forces" them apart, but rather because they were part of a big explosion a long time ago. This was not an explosion of the stuff in the universe into space, but an expansion of the space-time we described in Chapter 6.

I have found that many students have a hard time coming to understand Lemaître's explanation. For this reason, we will start with a simple-but-wrong model of the stuff of the universe exploding into space, and then show how it fits some, but not all, of the data. Then we will come back and do it right in Chapter 11.

I know it can be frustrating to spend time learning something you know is going to be wrong, but in this case, I have found it to be worth it. Since it turns out that the expansion of our universe is much more complicated than the explosions we see here on Earth, we will start with a two-step explanation. First, we will explain more about what explosions are like here on Earth. Then we will focus on the difference between explosions we are used to and what actually has been happening in our universe over the last fourteen billion years.

A simple explosion here on Earth can be thought of as a footrace. Imagine you are the race starter standing at the starting line. After your gun goes off, from your viewpoint, all the runners are moving away from you. It does not take too long before it is evident that the runners who are farthest away from you are the ones who run the most swiftly. This is what the data about the galaxies looks like: all the galaxies are moving away from us, with the farthest galaxies receding from our view the quickest.

While the real explanation is more sophisticated, the data we have presented so far is consistent with this simple picture of a universe that has galaxies racing away from us. This simple-but-wrong description of the universe as an explosion is shown in Figure 10.2. At any snapshot in time, the stuff that is farther away from the center of the explosion got there because it moved away faster. If we consider a race where the runners go off in all directions, we have a vision of why everything looks the same on all sides.

If we run the cosmic race backward in time—just like watching a film of the footrace in reverse—all of the galaxies would come together at a particular moment in the history of our universe. This would be its starting point.

The bottom part of Figure 10.2 shows the snapshots in reverse order—backward in time. If galaxies are moving away from us with faster speeds at greater distances, there must have been a single moment in time—a beginning—when everything was concentrated in one single point in space. This is where the explosion occurred; this was what has been called the big bang.

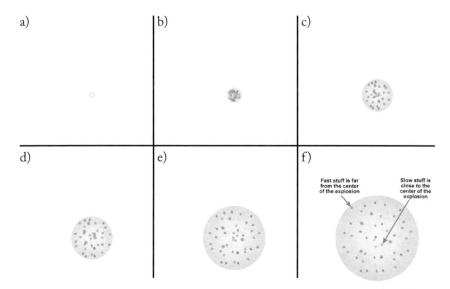

FIGURE 10.2 A simple-but-wrong explanation of the data from the galaxies. If we consider a basic explosion (shown here in six snapshots in time), then the fast stuff that comes from the explosion will have traveled farther; the slow stuff will be closer to the center of the explosion. In the middle row, we run the pictures backward in time (in reverse order). Eventually, we would get a snapshot that would be the explosion itself.

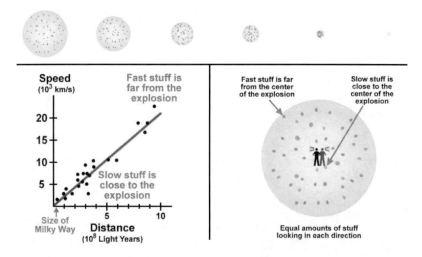

10.3 THE STORY IS MORE COMPLICATED...

Many non-scientists like the idea of a universe that began in a big bang because of its resemblance to the creation story described in the Bible. To scientists, a universe that begins with an explosion is appealing because it is both simple and

BIG BANG, BLACK HOLES, NO MATH

accounts for all the data we have just presented: All of the light from the distant galaxies is red-shifted; there appears to be an equal amount of galaxies on all sides of us; and the farther away the galaxy is, the faster it is moving.

The problem with this simple explanation is that it is based on the assumption that we are located at the center of the explosion, and that we are not moving; we are located at the race-starter position. This is troubling because the Earth isn't the center of the Solar System, and the Sun isn't even at the center of the Milky Way. Could we really be at the center of the universe? While this explanation is consistent with the data we have presented so far, perhaps we should explore some similar hypotheses.

In Figure 10.3, we consider two other simple and similar hypotheses to see if they are consistent with the data.

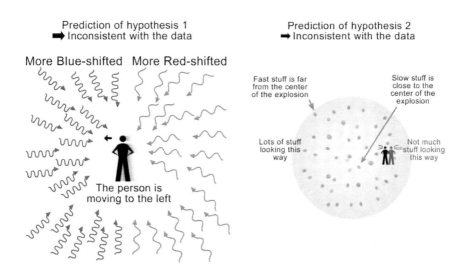

FIGURE 10.3 In this figure, we consider the implications of two competing—and wrong—hypotheses. In the first case (hypothesis 1), we consider the idea that we are at the center of the universe and are moving to the right. If this were correct, the data would look like the figure on the left where the spectral line shifts from the various galaxies look different in different directions. This is not what we observe. In hypothesis 2, we consider the case where we are "part" of the explosion and have traveled away from the center. In this case, the data would look like the figure on the right; there would be different amounts of stuff on each side. Also, we would see differing amounts of red-shift on both sides. This also is not what we observe. Neither of these hypotheses can explain our data.

In our first hypothesis, we consider the case where we are at the center of the universe, but are moving to the right. We quickly determine, however, that this cannot be a correct description because if it were true, the spectral line shifts from the various galaxies would look different in different directions. This is not what we observe in the data.

In our second hypothesis, we consider the case where we are not at the center of the universe, but rather part of the explosion; we are one of the runners in the race and are watching the other runners on all sides of us. This sounds more reasonable since, as far as we can tell, the Earth does not seem to be at the center of anything.

However, this hypothesis is also inconsistent with the data. If this were happening in nature, we should observe different amounts of stuff in every direction we look. We would also see varying red shifts of the galaxies in different directions: smaller in the direction we are moving, and larger away from the direction we are moving. But this is not what we observe. Neither of these simple hypotheses can therefore explain our data.

So are we really at the center of the universe and not moving? Is there a better explanation? This is the question we will address in the next chapter as we consider Lemaître's description. Like him, we will look to general relativity for a solution and explore the possibility that our universe is not an explosion *into* space, but rather an expansion *of* space (or, more correctly, space-time). When we get to Chapter 12, we will see even more evidence that supports the expansion-of-space explanation.

Expanding Space-Time CHAPTER

Based on the observations of galaxies we discussed in the last chapter, we can construct the simple-but-wrong hypothesis that the universe began by exploding into space and has been expanding ever since. However, this model only fits the presented data if we, as observers, are both at the center of the explosion and stationary. Evidence from Chapter 2 clearly shows that the Earth is moving and that it is neither the center of our own Solar System nor near the center of our galaxy. Although it is possible that we are moving slowly enough and are close enough to the center of the universe for the data to look this way, the fact remains that it seems unlikely.

In this chapter, we see how general relativity comes to the rescue by presenting a far better explanation for this evidence and more: we live in a universe with expanding space-time. Since this brain-warping idea is a foreign one for many of us, we will discuss how the expanding space-time model describes the data.

With this model in hand, we focus the next chapter on photons and hydrogen in the universe and how they provide powerful, smoking-gun evidence that space-time has been rapidly expanding for billions of years.

11.1 GENERAL RELATIVITY AND EXPANDING SPACE-TIME

The modern understanding of the universe's expansion, first envisioned by Lemaître, is not as an explosion *into* space, but rather an expansion *of* space-time. This might seem weird, but we have encountered this idea before on a smaller scale. In Chapter 6, we learned that general relativity describes how masses like our Sun curve space-time. Since general relativity states that space-time can change and curve, this means that space can also expand or stretch.

Because most students have a hard time visualizing an expanding three-dimensional space, I find it easier to instead have them imagine an expanding two-dimensional space. An example like our universe is shown in Figure 11.1, where we show the surface of an expanding balloon with galaxies drawn on it. Some people prefer thinking about a globe where the continents and islands are envisioned as galaxies.

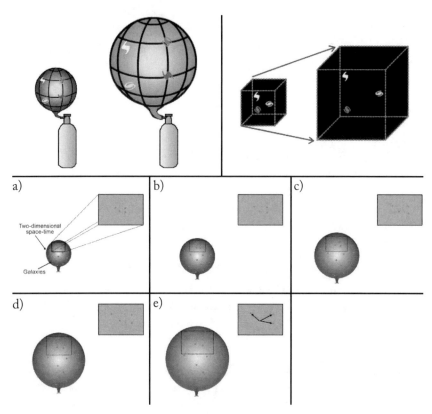

FIGURE 11.1 The expansion of space-time. In the top left, we show an analogy for two-dimensional space expansion. If we consider only movement on the surface of the balloon, it is equivalent of just being able to move in two dimensions. As the balloon gets bigger, "space" gets bigger. As the balloon expands, everything expands away from everything else. There is no place on the surface that can be seen as the center of expansion. Said differently, everywhere on the surface can equally be considered the center of expansion. In the top right, even if the galaxies are not moving through space, the points get farther from each other because space is expanding. In the bottom two rows, we show greater detail on how the points separate more quickly as they grow farther apart due to the balloon's expansion. Note that the distance between points is larger in the later figures. This is the same as a faster speed.

While a balloon is three-dimensional, here, we are only visualizing the *surface* of the balloon; moving only on the surface is the equivalent of being able to move in only two dimensions. For now, this is what we will call two-dimensional "space." As the balloon inflates, the surface gets bigger, so our space enlarges. Even though the drawings themselves don't move on the balloon, they all move away from each other as the balloon expands.

What makes the balloon example particularly helpful is that the surface of a sphere is a type of space with no center, in much the same way that there is no center of the universe. As the balloon expands, everything expands away from everything else. There is no place on the surface that is the center of expansion. Or, if you like, *all* places can be thought of as the center of the expansion.

Just like the surface of a balloon, from our vantage point, it looks like we are at the center of the expansion. In fact, it looks that way to anyone, anywhere in our universe. You can now see why we did not draw a three-dimensional expansion: it would mean describing the surface of a four-dimensional object, which no one can draw.

Another peculiar concept to note is that if we consider just the surface, there is no "inside" of the balloon. There is a center of the balloon, to be sure, but this is not on the surface. Also keep in mind that while the drawings on the balloon don't change their locations, galaxies can move through space, even if space itself is expanding.

So how does expanding space-time explain the different pieces of evidence? In an expanding space-time, there are two things that affect the speed of galaxies: their speed through space and how fast space is expanding.

Like a boat traveling on a river, even if the engines are off, the boat is carried along by the river. If the engines are on, it can travel against the flow of the river, or it can go with the river and move even faster. This is the same way that the galaxies zoom away from us at very high speeds (although galaxies do not have engines, and they aren't slowed down like if they were on the water since they just travel through space). As the universe expands, space itself expands and carries the galaxies along with it as shown on the right-hand-side of Figure 11.1.

Looking back to the data in Figure 10.1, we can think of the line as being the speed of the river. We can further think of each galaxy that happens to be traveling through space toward us as traveling with a smaller-than-average speed, and those galaxies that are headed away have an above-average speed. Switching to the balloon in Figure 11.1, we can see that even if the two galaxies are slowly moving toward each other (like marker dots on the balloon itself, if marker dots could move on a balloon), the expansion of the balloon (space itself) can make the distance between any two points increase.

If you think of a nearby galaxy, like Andromeda, it is being carried away from the Milky Way in the flow of space-time. However, it is moving through space toward us so quickly that it is expected to collide with us in about four billion years. On the other hand, if we have two galaxies that are very far apart from each other, but are moving through space toward each other, they can actually be moving apart. This also explains why, overall, distant galaxies move away more rapidly than nearby galaxies.

Since it is difficult to draw on a piece of paper, but is possible (albeit also hard) with a balloon, I encourage you to get a balloon, draw some dots on it, and then inflate it. You will easily see that as the size of the balloon increases, the distance between the dots also increases (see the bottom of Figure 11.1). What is harder to see is that the farther apart the points are on the balloon, the faster they separate as the balloon expands.

Now that we have discussed the expansion of space-time, we can take another look at the red-shift data from the distant galaxies. In the last chapter, we suggested that the red-shifts occur because galaxies are actually moving very quickly through space. We now have a different understanding, which is that the galaxies are not moving very quickly through space (should only give small red-shifts), but rather that space is expanding and stretching the wavelength of photons that travel through it.

According to general relativity, space can expand, which stretches a photon's wavelength. This means that at the same time a photon is traveling across an expanding universe, its wavelength becomes longer, or red-shifted.

The way we now understand the red-shifts of the light from galaxies is shown in Figure 11.2. In it, we see a star in a distant galaxy that emitted an infrared photon (10^{-5} m—the distance across a red blood cell) toward our galaxy billions of years ago. Space expands as the photon moves through it (the grid gets bigger), which stretches the photon to longer wavelengths. The farther away the galaxy of its origin and the longer it has traveled, the more red-shifted a photon becomes. By the time this particular photon arrives at our eyes billions of years later, the universe has expanded so much that the photon has stretched to a microwave wavelength (10^{-2}—the distance across a dime).

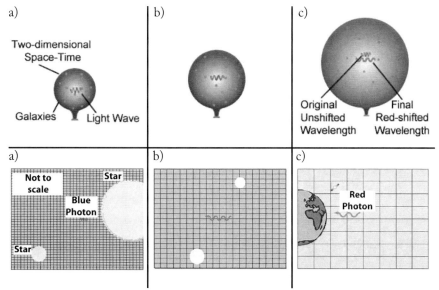

FIGURE 11.2 Six snapshots in time of a photon in expanding space-time. In the top three, we see how the wavelength stretches as the universe stretches. In the bottom set, we see light from a star in a distant galaxy on its path toward the Earth. The photon was emitted from the star on the right as a blue photon. In the middle plot, we see the photon as it travels through space and past other galaxies; it has traveled for so long already that the expansion of space-time has significantly stretched its wavelength. At even later times, the photon arrives at our eyes on the Earth having been stretched into a red photon. This stretching of the wavelength of light in an expanding universe explains the red-shifts of light from distant galaxies. It will also help explain the cosmic background radiation in Chapter 12. Note that the grid marks are not to scale. (Color version on page C-8.)

11.2 The Cosmological Principle and Why There is No Center of the Universe

The idea that space-time is expanding is consistent with both the data and the theory of general relativity. It explains how the data can make it "look" like we—or anyone else who might be out there—are at the center of the universe, even though we're not even at the center of our Solar System or galaxy. Essentially, we do not have to be in a special place in the universe for this data to make sense.

It is important to point out that we cannot *completely* test the idea that the expansion looks the same to everyone because we cannot be everywhere out there to check. The idea that the universe looks (basically) the same to *everyone, everywhere* in the universe, is called the **cosmological principle**. Although we cannot prove it is true, it is consistent with all the known data.

In the next chapter, we will describe photons and hydrogen in the universe and what happened to them as the universe changed over time. Both will provide strong evidence for the expanding space-time model. As we said before, in many ways, they are the most compelling evidence for the big bang—or at least that the universe was once much smaller and hotter than it is today—but they are hardest to explain to the non-scientist.

Now that we have discussed expanding space-time, we are ready.

Photons and Hydrogen in the Universe

CHAPTER 12

In this chapter, we explore two additional—and completely different—pieces of evidence that support the big bang theory. These two pieces of data provide, perhaps, the most compelling evidence that we live in a universe that was once very hot and dense and has been experiencing expanding space-time ever since. We start by sharing these pieces of evidence and then laying out an abridged version of the history of the universe to see where they fit in.

The first piece of evidence is the enormous number of low-energy photons that permeate the universe, also known as the **cosmic background radiation**. There are about four hundred photons per cubic centimeter throughout the entirety of space. They are equally distributed in all directions and have a temperature of about 2.7 degrees above absolute zero (Kelvin). It is as if they are, or were, in thermal equilibrium.

The second piece of evidence is that about 91 percent of the atoms in the universe are hydrogen, while most of the other 9 percent are helium. There are precious few atoms of any other type, except for those collected in small places like Earth.

To understand why these two facts provide such compelling evidence, we must understand that the way the universe changes over time—what physicists call the **evolution of the universe**[1]—is essentially the story of two important facts: (1) as the cosmos expands, the energy of the particles in the universe drop; and (2) the *way* that particles interact with each other critically depends on the energy of the particles. Taken together, our pieces of evidence are smoking-gun clues for the big bang theory.

---

[1] Not to be confused with the way that species of animals on Earth evolve over time.

12.1 THE PARTICLES OF THE EARLY UNIVERSE, THEN AND NOW

We begin our description of the history of the universe a mere second after the big bang, and then explain why the universe looked the way it did back then. We will refer to this time as the **early universe**.

Right after the bang, the cosmos was much smaller than it is today, and there were no galaxies—just lots of particles moving in space. Using a previous analogy, all the particles in the universe were confined to the surface of a mini-balloon. Since space was much smaller then, all the wavelengths of the particles would be much smaller than they are today, thus all the particles would be high-energy.

By this time in our history, the different types of particles—like protons, neutrons, electrons, photons, and others mentioned in Chapter 3—would have interacted with each other, created lots of other particles (see Figure 9.7), and effectively come into thermal equilibrium (see the left side of Figure 12.1). We would describe the early universe as being small and as having a high temperature.

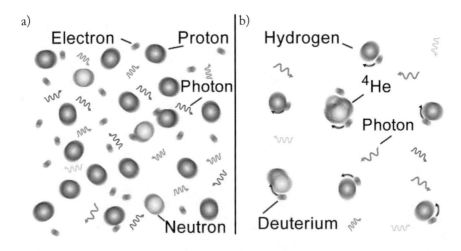

FIGURE 12.1 The photons, electrons, protons, and neutrons in the early universe "a" and today "b." In the early universe, the particles were free, and had a very high temperature. Today, the temperature is much lower, and all the protons, neutrons, and electrons are in atoms. The photons still travel freely.

As time goes by, the universe expands and the particles both interact and decay in different ways. The unstable particles—such as neutrons outside a nucleus—decay away into electrons, protons, and neutrinos. Since there is no reason to believe that the protons and electrons should have all vanished (they do not transform into something else, evaporate, or disappear), most of them are still around today; the majority are hydrogen atoms and many are in stars. We will explain why they are primarily hydrogen soon.

Time marches on and the space in the universe is expanding and stretching all the wavelengths of the particles, especially the photons. In other words, over time, the energy of the photons is dropping and the temperature of the universe is dropping. Since photons do not decay, the ones from the early universe are still out in space, as well. Indeed, they would have a low temperature distribution (low energy) today because they had a high temperature billions of years ago.

Sophisticated calculations, which take all the data into account, point to a universe that came into existence about fourteen billion years ago. This remarkable number agrees with the ages of other ancient things we observe. For example, the galaxies farthest from us emitted their light about thirteen billion years ago, and the age of the Earth—based in part on the dating of the oldest rocks discovered—are about 4.5 billion years old. If we found something more than fourteen billion years old, we would worry that we were missing something, or that our story was wrong. Remember: any one piece of solid evidence can invalidate an entire story.

12.2 WHAT HAPPENS WHEN PARTICLES INTERACT?

This short description of how the universe changed over time leaves many questions: Why do we think there were only particles in the early universe? How did they combine to form the stars and galaxies we now observe? Why are there lots of photons in the universe? And why are almost all atoms hydrogen atoms? An equally interesting question is: How did we get any heavy atoms, like gold or uranium, at all?

The answer to these questions comes from the fact that at different energies, *extremely* different things can happen in the interactions between particles. For example, while protons, nuclei, and atoms can be created in interactions, they can also be broken apart if they interact with a high-enough-energy particle. Depending on the energies of the particles, fundamental particles can combine to form composite particles, and composite particles can get broken up into fundamental particles. Add to this mix the fact that particles can decay, and things can be pretty complicated. These many factors all conspire to drastically alter what goes on in the universe as it gets older and colder.

Figures 12.2, 12.3, and 12.4 show examples of how particles combine. We see three quarks combining to form a proton, a proton and neutron combining to form a deuterium nucleus, and a proton and electron combining to form a hydrogen atom. Other possibilities are shown in Figures 8.3, 8.4, and 8.5.

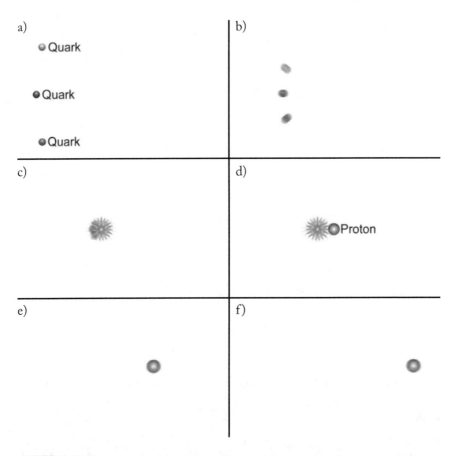

FIGURE 12.2 This figure, and the next two figures, shows a series of action shots showing particles combining in the early universe. Here we see a simplified view of three quarks combining to form a proton.

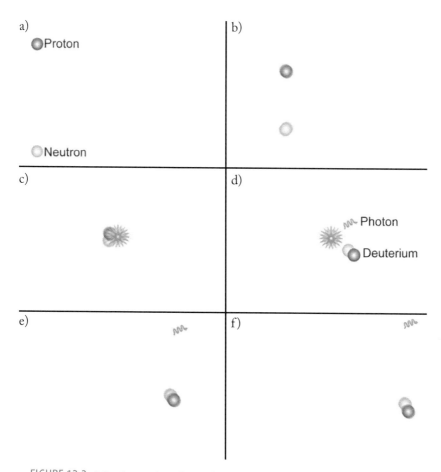

FIGURE 12.3 Like the previous figure, this set of pictures shows a proton and neutron combining to form a deuterium nucleus.

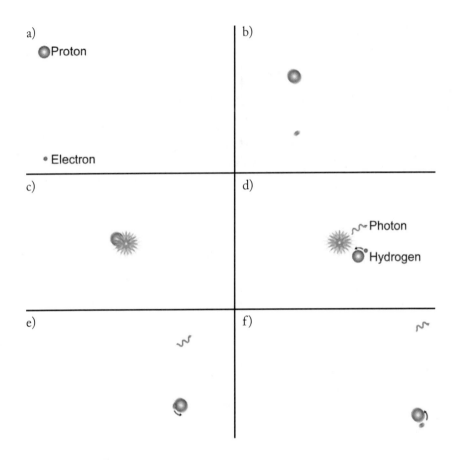

FIGURE 12.4 Like the previous two figures, this set of pictures shows an electron colliding with a proton to form a hydrogen atom. Note that the energies of the photons emitted in this figure are much lower than the energy in Figure 12.3.

Figure 12.5 shows four separate interactions with a photon—two in which the photon has high energy, and two where it has low energy. If the photon's energy is high enough, it can break the atom apart (into electrons and the nucleus) or the nucleus apart (into protons and neutrons). If the photon's energy is low, it will, at most, bump into the nucleus or atom and change its direction. It takes a much higher-energy photon to break apart a nucleus than it does to break apart an atom.

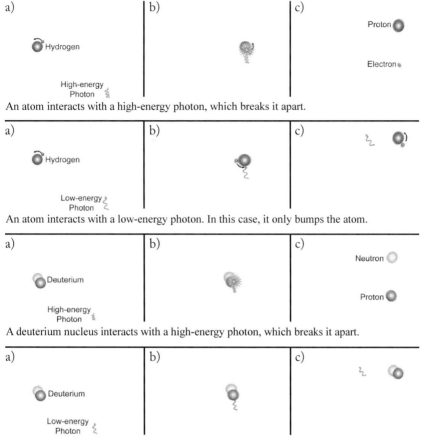

a) Hydrogen

High-energy Photon

b)

c) Proton Electron

An atom interacts with a high-energy photon, which breaks it apart.

a) Hydrogen

Low-energy Photon

b)

c)

An atom interacts with a low-energy photon. In this case, it only bumps the atom.

a) Deuterium

High-energy Photon

b)

c) Neutron Proton

A deuterium nucleus interacts with a high-energy photon, which breaks it apart.

a) Deuterium

Low-energy Photon

b)

c)

A deuterium nucleus interacts with a photon that does not have enough energy to break it apart. In this case, it only bumps the nucleus.

FIGURE 12.5 Action shots showing various interactions between atoms and nuclei with a high-energy (gamma ray) photon or a low-energy (microwave) photon. Note that the wavelengths of the photons are not drawn to scale.

There are many other types of interactions as well. For example, two high-energy photons can collide and turn into an electron and a positron (see Chapter 9). However, since mass and energy are equivalent ($E=mc^2$), this only occurs if the photons have an energy high enough to produce the mass of both electrons. If not, the photons will bump into each other and move off into space, or just ignore each other completely.

On the flip side, since electrons and positrons always have mass, they can interact to produce a pair of photons. This is important later in the story. We will see more types of interactions like this in Chapter 19. We are now ready to explain the particles in the hot, early universe and why there were no galaxies.

12.3 A HOT, EARLY UNIVERSE HAS LOTS OF PHOTONS AND HYDROGEN, BUT NO HEAVY NUCLEI

The enormous amount of high-energy particles in a small space quickly brought the early universe to have the same temperature everywhere. There are numerous reasons why so many high-energy photons, electrons, and light nuclei (mostly hydrogen) were present. For starters, even if whatever process that initiated the universe did not directly produce a large number of photons, collisions between electrons, positrons, or other particles would have soon produced them. Furthermore, even if heavy nuclei or atoms (or stars or galaxies) were part of the early universe, they would have quickly broken apart.

A hot universe filled with high-energy photons only has light nuclei. This explains why we do not have many heavy-nuclei atoms today. To understand why, consider what would have happened back then.

It is certainly true that in the early universe, there were lots of protons and electrons in a tiny space, so numerous hydrogen atoms would quickly form. However, since the early universe was chock-full of high-energy photons, it would not have taken long for any single hydrogen atom to encounter a high-energy photon. The collision would split the atom back into a proton and an electron (see Figure 12.5). This means there were essentially no atoms in the early universe, but rather large numbers of free electrons and protons.

Similarly, while deuterium can be produced when a proton and a neutron, or two protons, find each other (see Figures 8.3-8.5 and 12.3), it will also not last very long. Quickly after formation, the deuterium would encounter a high-energy photon, which would break it apart (see Figure 12.6). Similar arguments hold for any nuclei, although some nuclei are harder to break apart than others. So even if there were copious quantities of heavy nuclei or atoms in the early universe, the high-energy photons would soon destroy them.

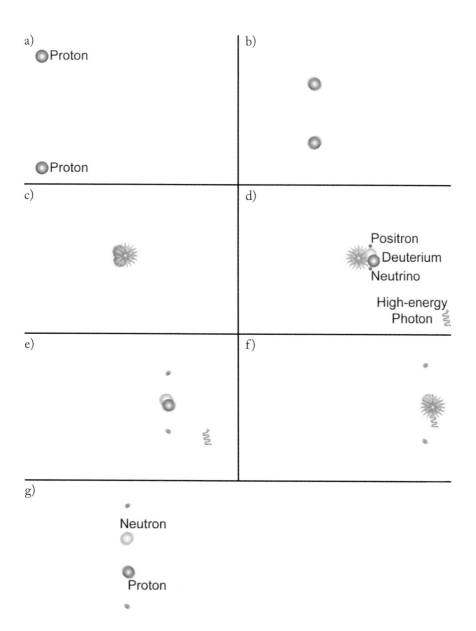

a)
Proton
Proton

b)

c)

d)
Positron
Deuterium
Neutrino
High-energy
Photon

e)

f)

g)
Neutron
Proton

FIGURE 12.6 High-energy photons break up nuclei before they can either find electrons to form atoms or find other nucleons to form heavier nuclei. In this case a high-energy photon breaks apart a deuterium nucleus.

At mere seconds old, the early universe contains only photons, electrons, positrons, protons, neutrons, and neutrinos. Again, we note that when neutrons are not part of a nucleus, they will eventually decay. But at this early point in the universe's history, such an occurrence is rare.

As the universe ages and the average energy of its particles decreases from spacetime stretching the wavelengths, the way in which particles interact changes drastically. When the universe is about three minutes old (three minutes after the bang), the average photon no longer has enough energy to break apart nuclei; the interactions resemble those shown in Figure 12.5 for low-energy photons. Since deuterium can still be produced (see Figures 12.3 and 12.6), but is not always destroyed, it is just a matter of time before it becomes more plentiful. More deuterium means that we do not have to wait too long before a deuterium nucleus interacts with a proton to form helium (^{3}He, using the notation from Chapter 8).

The creation of helium, however, is but one of many possible outcomes from interactions involving deuterium. Two deuterium nuclei, for instance, can fuse and form a ^{3}H and a proton. As the universe ages, it will not take long before a fair amount of ^{4}He is produced in these same types of processes.

Despite large numbers of ^{4}He, nuclei heavier than helium will not accumulate in the universe. This can be understood from studies of nuclei interacting in laboratory experiments. Scientists can form nuclei that are heavier than helium just by adding protons or neutrons to helium; however, these nuclei are not stable, each quickly decays back into helium (more on this in Chapter 14).

Essentially, in order to get heavy nuclei, you need lots of helium in a small space (like in a star). Since the universe is rapidly expanding, it takes a while to build up even a slight amount of helium. Thus, in the early universe, the creation of nuclei that are heavier than helium is infrequent. Most of these nuclei (carbon, oxygen, iron, etc.) will be created in stars hundreds of millions of years later. We will discuss this more in Chapter 16.

The bottom line is that a hot, early universe in thermal equilibrium explains the number of each type of atom in the universe: mostly hydrogen, a little helium, and very little of any other type of atom.

At this point, we will stop talking about atoms in the universe and focus on photons. We will return to atoms in the next section, when we discuss a more detailed history of the universe. This will include a look at the formation of atoms a few hundred thousand years after the bang. Then, in Chapters 15 through 17, we will consider how atoms condensed into galaxies, stars, and black holes about half a billion years after the bang.

12.4 Photons in a Cold Universe

Photons play a very different role in the early universe than they do now, in its later years. In the early universe, a high-energy photon quickly breaks apart any heavy nucleus or atom that is created. Eventually, however, the temperature of the universe drops enough that the photons no longer have adequate energy to regularly knock electrons out of their atomic orbits. This occurs when the cosmos is several hundred thousand years old and has a temperature of about 3,000 Kelvin. At this point, the average photon wavelength corresponds to visible light (see Figure 12.7).

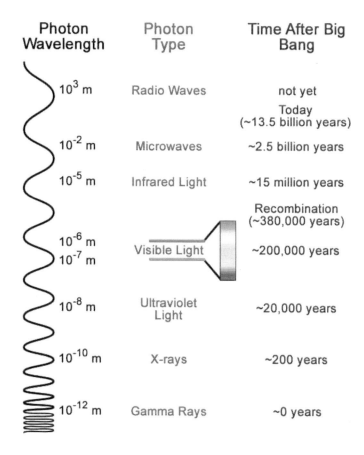

Photon Wavelength	Photon Type	Time After Big Bang
10^3 m	Radio Waves	not yet
		Today (~13.5 billion years)
10^{-2} m	Microwaves	~2.5 billion years
10^{-5} m	Infrared Light	~15 million years
		Recombination (~380,000 years)
10^{-6} m 10^{-7} m	Visible Light	~200,000 years
10^{-8} m	Ultraviolet Light	~20,000 years
10^{-10} m	X-rays	~200 years
10^{-12} m	Gamma Rays	~0 years

FIGURE 12.7 A photon's wavelength stretches as the universe expands. The photons in the early universe were high-energy gamma rays. Now they are low-energy microwaves.
(Color version on page C-7.)

The times after photons no longer interact with atoms—or much of anything else because of quantum mechanics (see Chapter 7, Figure 7.7 in particular)—are very different than before. We will talk about this more in Chapter 14. Until then, we simply note that now if a photon can't knock an electron out of an atom, for the most part, they travel freely through space, largely ignoring any atom it does encounter. Since photons are stable, they can travel forever and do so in a universe that is expanding and transforming from a high-temperature to a low-temperature place.

So around four hundred thousand years after the bang, photons had stretched so much that they stopped interacting with atoms. Since space today is about a thousand times bigger, the wavelength of the average photon is about a thousand times longer. The visible photons, then, are microwave photons today (see Figure 12.7). These low-energy photons that fill our low-temperature universe are hard to find, but were uncovered in 1964 by Arno Penzias and Robert Wilson.[2] This discovery convinced most scientists of the validity of the big bang model.

As mentioned earlier in this chapter, low-energy microwave photons are often known as the cosmic background radiation. It is useful to define the words in this phrase. "Cosmic" means something relating to the cosmos, or something that it is everywhere in the universe. The dictionary defines "background" as being "inconspicuous; out of sight or notice." And "radiation" is "energy in the form of waves or particles" (photons are a type of radiation). So our phrase is a collection of fancy words that basically means there are many low-energy photons throughout the universe, which makes them difficult to notice. Instead of saying "cosmic background radiation," most scientists use the phrase **cosmic microwave background (CMB),** since the photons are in the microwave range.

These photons are important not only because of their mere existence, or even because of their vast numbers. Rather, their full importance can be understood from their energy and how they are distributed in space. The data shows that the energy distribution is consistent with a specific temperature. Equally important, the temperature of these photons is virtually the same in every direction. How uniform is the temperature of the cosmic background radiation? If we look at the entire sky in a single map and stretch out the sphere onto a flat page, we can draw a temperature map where different colors correspond to different temperatures. As shown in Figure 12.8, the cosmic background radiation is incredibly uniform, but not perfectly so. We will come back to that in Chapter 15.

---

[2] This is a wonderful story and I encourage you to read about it using the texts in the Suggested Readings. One of the things that makes it so much fun is that Penzias and Wilson were not looking for these photons; they stumbled upon them while they were working on other things and had no idea what the "problem" with their experiment was. For awhile, they thought it might be due to pigeon poop in their detector. It was not until they were talking to some physicists who had been looking for the photons that the pair realized what they had discovered. They got the Nobel Prize; their colleagues did not.

NASA / WMAP Science Team

FIGURE 12.8 On the left is a two-dimensional representation of the Earth. Similarly, but looking out, is the temperature map of the cosmic background radiation. Different temperatures are shown with different shadings. As you can see, the incredibly uniform shading shows that the temperature is the same in every direction.

Perhaps most importantly, not only do the photons have an energy distribution consistent with a specific temperature, but also that temperature is really low. It is measured today to be 2.728 Kelvin, which is -270 degrees Celsius or -455 degrees Fahrenheit. All of this data is consistent with what you would see in an old, cold and expanding universe that had been in thermal equilibrium at one point in its history.

12.5 SUMMARY

The enormous amount of evidence gathered by scientists points to a universe that was once a very hot and tiny place and that had the same temperature everywhere. This same universe, the evidence shows, has been experiencing a space-time expansion described by general relativity for about fourteen billion years. Supporting data includes the large red-shifts of distant galaxies; the fact that the vast majority of atoms in the universe are hydrogen and a much smaller fraction are helium (and that we can predict the exact amounts); and the incredible number of photons that have an energy distribution consistent with a very low temperature that is virtually the same in all directions.

We observe a universe that appears to have been in thermal equilibrium, but now possesses a temperature that reflects fourteen billion years of aging. The expansion of space-time helps explain why galaxies are moving away from us in the same manner in all directions; why the more distant ones move more quickly; and why the background radiation both exists and looks the same in all directions. While there may be a better explanation out there that we do not yet know about, the big bang theory does a remarkable job of explaining an amazing amount of data. Otherwise, these separate pieces of evidence would appear to be an extraordinary coincidence.

Next, in Unit 4, we will learn more about the evolution of the universe. In particular, we will provide more detail about what happened right after the big bang, as well as follow the history of the universe from the first millionth of a second after its inception, through galaxy and black hole formation, and to the cosmos we know today.

The Evolution of the Universe:
What Happened After the Bang

In Unit 3, we described the hot, early universe and presented some of the evidence for the big bang theory. In this unit, we will tell a more detailed version of the history of the universe, from its earliest moments until today. If we were to be concise, we might say something like, "About fourteen billion years ago, the universe was a hot and tiny place and has been expanding and cooling ever since." Our tale, though, is far more interesting than that. We will tell it in two chapters:

> Chapter 13: The Early Universe
> Chapter 14: After the First Three Minutes

The central characters in the story are particle energies and universe size (or, better yet, density) at any given time. Both are important as they determine not only how particles interact, but also how often they find each other to do so. How likely, for example, are a proton and an electron to find each other and combine to form a hydrogen atom?

In some sense, the main theme in the history of the universe is how particles have interacted over time and how they formed the nuclei, atoms, stars, and galaxies that we see today.

The Early Universe  CHAPTER 13

The best place to start a story is clearly at the beginning. Unfortunately, there are a number of reasons we do not understand the "bang" part of the big bang—what caused it, what came before it, or even if there actually was a bang. For starters, general relativity cannot make predictions at infinitesimally small sizes, and quantum mechanics does not explain the curving of space-time. Also, as our universe appears to have been in thermal equilibrium at some stage, it is not obvious how to determine how it all began. We have a lot of ideas and data, but no conclusive evidence.

The best we can do, then, is to start a short time *after* the beginning. Instead of focusing on the birth of the universe, we describe its infant, toddler, and adolescent stages. The good news is that we are confident about this part of the story. From almost any time during the early universe, we can work our way backward in time, getting very close to the beginning itself. We can also work our way forward, and describe the history all the way to the universe we know today.

The big bang theory says that the universe started with a bang. Okay, then what? As stated in the previous chapter, the universe expands and stretches the wavelengths of the particles. This stretching lowers the particles' energy in the universe which, in turn, changes the way they interact (or collide) with each other.

While we had a brief overview in the last couple of chapters, we next undertake a more extensive overview. After that, we will scrutinize a detailed, step-by-step version of events, this time starting at a millionth of a second after the bang, to reveal a more complete and interesting story. We begin at a millionth of a second—and not before—simply because the story is easier to tell starting at that point. We will say more about times before a millionth of a second in Chapters 19 and 20.

13.1 A Brief History of the Universe

Sometimes a picture is worth a thousand words. Figure 13.1 shows a brief history of our universe with some of the important milestones along the way. What we see is that since the beginning, about fourteen billion years ago, the universe has been expanding and its temperature has been dropping. Over time the "stuff" in the universe interacts and forms nuclei, atoms, stars, and galaxies. The history of our universe can basically be described, then, as the story of particles knocking into each other as the universe gets colder.

As we saw in previous chapters, in a collision between particles, a number of things can happen. If there is enough energy, different types of particles can be created. Another possibility is that colliding particles combine to form a composite object such as a proton, nucleus, or atom. High-energy collisions can also go the other way and break these objects apart. Thus, which particles exist and what energies they possess has a huge impact on what happens at any point in time.

The history of the universe depends on how particles interact. This knowledge allows us to break the cosmic story down into a few short chapters that we sum up in Table 13.1. We are now ready to start with more detail in our story: the very early universe.

How long after the bang	What's happening
Zero	The big bang
One millionth of one second	Quarks combine to form protons and neutrons
A few minutes	Protons and neutrons combine to form deuterium and helium nuclei
A few hundred thousand years	Nuclei and electrons combine to form atoms
One hundred million to one billion years	Atoms combine to form stars and galaxies
Nine billion years	Our Solar System forms
~13.7 billion years	You read this book

TABLE 13.1 What is happening at various times during the history of our universe.

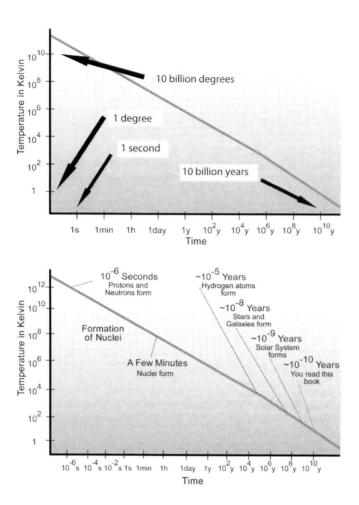

FIGURE 13.1 A simplified version of the history of the universe in two different time lines. As in Chapter 2, we use a graph to help describe the story. This is a different type of graph, however, in that it tells us the universe's temperature at various times in its history. To find out what the temperature is at 1 second after the bang, for instance, you would look to the horizontal axis and find the tick-mark labeled "1 second." Then you look up to the red line to see how "high up" it is. We see that it is slightly above the tick-mark labeled "10 billion Kelvin" on the vertical axis. Thus, at about a second after the bang, the temperature was slightly more than 10 billion Kelvin. Similarly, about 10 billion years after the bang, the temperature is slightly more than 1 Kelvin. As time goes by, the universe expands, stretching the wavelength of the particles in it, which lowers their energies. In other words, the temperature drops. As the temperature changes, the energies of the different types of particles affect which types of reactions can occur. Note, that while the red lines are drawn as straight here, in reality the drop in the temperature isn't quite as smooth.

13.2 THE VERY EARLY UNIVERSE

Approximately one millionth of a second after the bang, the universe was very dense and all the particles had short wavelengths (high energy). Thus, the universe was insanely hot (more than a trillion degrees), and there were large numbers of high-energy particles moving around; the universe had already come to have the same temperature in all directions (and may have actually been in thermal equilibrium).

Today, most protons and electrons are parts of atoms which, in turn, are parts of galaxies, stars, and planets. But in the very early universe (before 25 microseconds), there were no composite particles. Instead, quarks, photons, electrons, and all other fundamental particles moved about independently, as shown in Figure 13.2.

It is not hard to understand why this is so. While composite particles could have been created directly in the big bang—or by the combining of fundamental particles—back then, any composite particle that did exist would quickly encounter another high-energy particle. In that collision, the composite particle would be broken apart. This is especially true in the very early universe when the cosmos was denser; particles were closer to each other and more likely to collide than they are now.

No matter how it started, the hot, early universe would eventually be filled with a vast array of fundamental particles, many of which are no longer commonly encountered.

While we have already talked about electrons and positrons colliding to produce two photons (Chapter 9), and vice versa, there are many other possibilities. For example, a collision between two high-energy photons could create a muon and an anti-muon, which are heavier cousins of the electron; we will talk more about them in Chapter 19. In fact, the high-energy collisions back then would have produced all the particles listed in Table 3.1.

Today, the only places we can find such exotic particles are when they are produced by the world's highest energy accelerators. Two of the most well-known are Tevatron, located at Fermilab outside of Chicago; and the Large Hadron Collider (LHC), located at CERN, which is in—or, more accurately, under—Geneva, Switzerland.

As their name suggests, particle accelerators speed up particles to very high energies, then point them on a collision path with each other. Essentially, they are not recreating the big bang, but they are recreating what the universe looked like in its earliest moments.

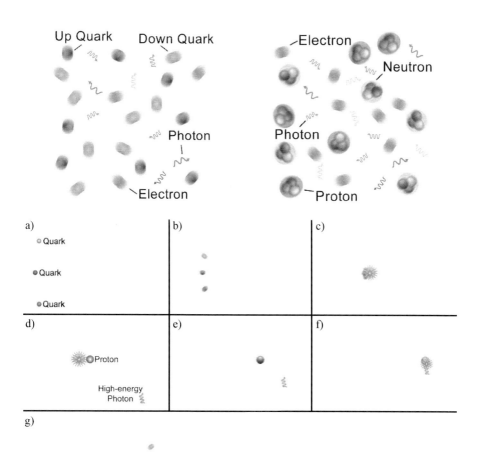

FIGURE 13.2 Particles in the very early universe. Above left: Before a millionth of a second after the bang, quarks and other particles move about in thermal equilibrium without combining. Above right: Following a millionth of a second after the bang, the photons are not energetic enough to break apart protons, so we have a universe filled with protons and neutrons. Below: Before a millionth of a second, quarks can combine to create protons, but because there are so many high-energy photons in the universe, they are quickly broken apart.

Getting back to the universe, since muons *can* exist and be produced in a high-energy collision, they could and would have been produced in the high-energy collisions of the early universe. Figure 13.3 shows them in action with other particles in thermal equilibrium. Since any particle that can exist *will* eventually be produced in a collision, a complete depiction of the very early universe becomes very complicated indeed. A proper description requires complete knowledge about what particles can exist in nature and how they interact with all the other particles.

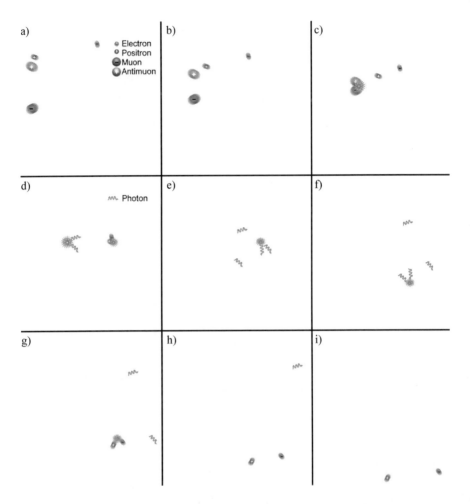

FIGURE 13.3 In the early universe, the energy of the particles is high enough to create lots of unstable particles, like muons, in thermal equilibrium. Each can interact and produce photons which can, in turn, produce electrons and positrons.

Unfortunately, while scientists have studied many particles in great detail, we do not know if we have discovered *all* of the fundamental particles. In fact, we have good reason to believe that there are still undiscovered particles out there. If we are right, we need bigger particle accelerators and/or other tools to detect them.

Since we know the fate of all the particles listed in Table 3.1—they are either stable (living forever) or unstable (decaying after some amount of time)—we understand why most of them were present in the early universe, but are no longer around today. By a millionth of a second, most would have decayed into everyday particles, such as electrons, photons, and protons.[1] Said differently, the unstable ones have decayed into stable particles, so only stable particles remain.

Figure 13.4 models the particles in our universe as water in a bathtub. The amount of water in the tub represents the quantity of a particular particle type in the universe. We have one tub for protons, one for electrons, one for photons, etc. In our analogy, high-energy collisions that produce particles are like the faucet pouring more water into the tub. Similarly, the drain symbolizes particles decaying or colliding and turning into other particles.[2] The battle between the faucet and the drain determines how much water is in the tub.

In other words, the energies of the particles—and how the particles interact when they have that energy—determines how much of each type of particle exists in the universe at any given time in its history.

Let us focus on protons as an initial example.

In the early universe, quarks collide and combine to form protons, thereby increasing the number of protons in the universe; this is the faucet in the figure. Before a millionth of a second, there are ample high-energy photons that can break apart any proton created—thus the drain is sucking away protons faster than the faucet fills the tub. As shown in the figure for this case, there will not be much water in the tub.

As the universe gets older and colder, the average energy of its photons drops low enough that the protons are no longer broken apart. Essentially, the drain in the tub gets clogged and the water level rises. This cannot keep happening forever. Eventually, all the free quarks will be used up, no more are produced, and the faucet is turned off. At this point, the tub is as full as it will ever be.

The tub analogy holds for other particles, as well. A second example, muons, is also shown in Figure 13.4. In the early universe, there is some water in the tub as high-energy collisions produce muons (the faucet), but the tub never gets too full because the muons are constantly decaying and/or colliding (the drain). As time goes by, however, the temperature of the universe drops and creation of muons ceases, but their decay continues. By around a millionth of a second after the big bang, virtually nothing can produce muons anymore, and soon thereafter, they will all have decayed away.

---

[1] Neutrons are more complicated since they decay after awhile if they are by themselves, but are "stable" and live forever if they are inside a nucleus.

[2] In essence, this would provide a faucet for a different type of particle.

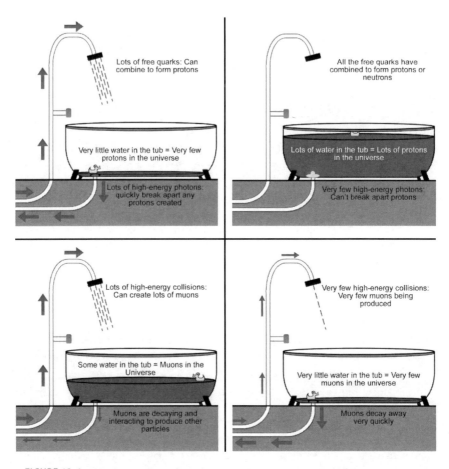

FIGURE 13.4 Water in bathtubs that represent the number of particles in the universe. The water from the faucet represents particles being created, and water going down the drain represents them decaying or changing into something else in a collision. The top left shows the number of protons before a millionth of a second after the bang; lots of quarks combine to become protons, but they keep getting busted apart in the early universe. Thus, there are not many protons in the universe—and not much water in the tub. The top right shows a later time when the universe is not as hot, so the photons no longer break up the protons and the tub fills. The bottom row shows the same thing, but for muons at early and later times.

We can now better understand why we started our discussion at about a millionth of a second after the big bang. At this time, the universe possessed a temperature in which the average energy of the particles was both too low to create more unstable particles and too low to continue breaking apart protons and neutrons. In addition, virtually all of the unstable particles, like muons, had decayed away into stable particles like protons, electrons, photons, etc. It is not that this time period is of particular importance; it's simply an easy time to describe our universe.

One more thing to recognize before we leave this early time is that it is not known how or when dark matter was produced in the universe. Scientists think there is a good chance it is a type of particle. They are using colliders, like the Tevatron or the LHC, to try to recreate particle collisions of the same energy that existed in the early universe. If they can do so, then they can study the particles and better understand when dark matter was created in our universe's history. For now, though, we will leave dark matter out of the story and come back to it later.

13.3 A Hundredth of a Second After the Bang, and Ten Seconds After the Bang

We pick up the story at about a hundredth of a second (0.01 seconds or 10^{-2} seconds) after the big bang. The universe has expanded and cooled significantly. It is now only about a hundred billion degrees. There are no unstable particles left, and all the quarks are contained in protons and neutrons.

We might guess that the number of atoms and helium nuclei would be rising. With all these protons and neutrons in the early universe, it is natural to think they would quickly find each other and fuse. Similarly, protons and electrons ought to combine to form atoms.

While deuterium and helium nuclei, as well as atoms, are quickly forming, it turns out that the overall quantity of each (the amount of water in the tub) remains small. This is because the universe is densely filled with energetic photons. Every time protons and neutrons combine to form a deuterium nucleus, the nucleus quickly encounters a high-energy photon or electron that bursts it apart (see Figure 12.5).

Similarly, a high-energy photon can break apart any proton and electron that come together to form a hydrogen atom. It will therefore be awhile before the universe is cool enough for atoms and heavy nuclei to stick around very long. The energy of the particles in the universe is still high enough for the production of lots of positrons, however, so many of those remain.

About ten seconds after the bang, the universe changes dramatically. At this time, it is a mere three billion degrees—low enough that photons can no longer break nuclei apart. Nuclei such as deuterium can now be produced and survive.

While it also is now cool enough for helium nuclei to form, it will take awhile before enough deuterium builds up so they can regularly do so. Electrons and positrons can interact and annihilate each other, but photon pairs are no longer energetic enough to turn into particle pairs, as shown in Figure 13.5. This means the number of electrons and positrons is dropping rapidly.

It is still too hot for atoms to stay together for any length of time. This will remain the case for the next couple hundred thousand years, although neutrons outside a nucleus will decay away by then.

Before we end this chapter, it is worth saying a few words about anti-matter particles in the universe.

One of the things we observe today is that there are many more particles than anti-particles. For example, there are enormous numbers of electrons in the universe, mostly in atoms, but almost no positrons.

Despite years of trying, we still do not understand why this is the case. We do know, however, that in very high-energy collisions, we see equal numbers of anti-matter and matter produced. The answer to this important puzzle, then, is likely to reside in the subtle ways that particles decay. For now, we simply accept the observed fact that there are many more electrons than positrons in the universe and pick up our story in the next chapter.

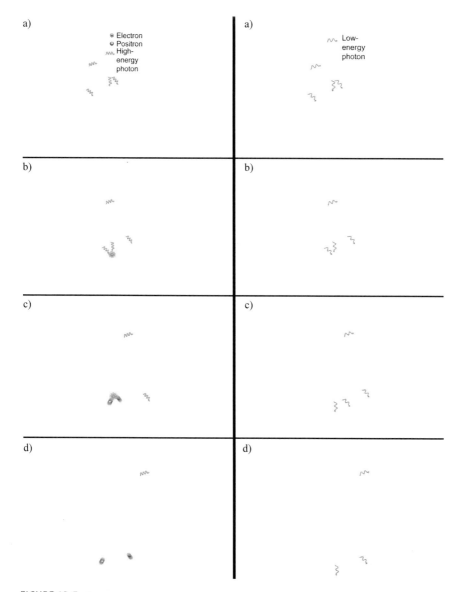

FIGURE 13.5 Photons with high and low energies. At very early times after the bang, the average photon is energetic enough to produce pairs of particles and anti-particles, as seen on the left. The picture to the right shows later times, when the photons still interact, but are not energetic enough to produce any type of particle. When this happens, the numbers in the universe of that particle type drops because they can freely annihilate each other, but cannot produce new particles. At a millionth of a second after the bang, muon pairs are rarely produced anymore. At ten seconds after the bang, photons are no longer energetic enough, on average, to create electrons and positrons. After these times, the number of each of these particle types quickly drops.

After The First Three Minutes

CHAPTER 14

So much happened in the first three minutes after the bang that it took a whole chapter to describe the events. Now, in fewer pages, we look at what happened over the next fourteen billion years. Despite a deceptive chapter length, our cosmos goes through some profound changes during this time period (see Figure 14.1).

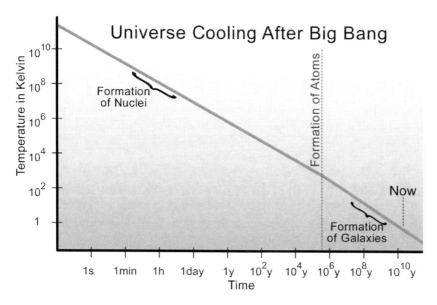

FIGURE 14.1 A simple history of the universe using the same time line and graph ideas as in Figure 13.1, but with different markers. As time goes by, the temperature drops and various types of particle reactions can occur. Here we highlight the formation of nuclei, atoms, and galaxies in the history of the universe.

14.1 THREE MINUTES AFTER THE BANG

At the three-minute mark after the bang, the temperature of the universe is about a billion degrees. This means that the average energy of the photons is now low enough that they cannot break apart a nucleus, but is still too high for any atoms that are created to stay together. Thus, since deuterium nuclei can regularly form (like in Figures 8.3-8.5 and 12.2), the amount of deuterium in the universe is rising. Similarly, from this point on, it is easier to build up heavier nuclei like ^{3}He and ^{4}He. Since all of these nuclei are stable, they can exist forever.

While it is natural to think that nuclei heavier than ^{4}He should start accumulating in the universe from collisions, this does not happen. It certainly is true that there is plenty of ^{4}He, and it does not take long for one to encounter a proton. However, the electric charge repulsion between the helium and the proton is so great that they often repel each other and very little lithium (^{5}Li, to be exact) forms.

In addition, experiments here on Earth show that ^{5}Li is not stable. Thus, every time a heavier nucleus like ^{5}Li is created, it is more likely to decay than it is to find another neutron to make ^{6}Li (see Figure 14.2).

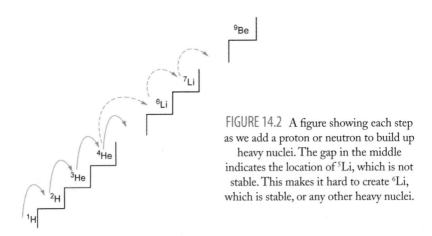

FIGURE 14.2 A figure showing each step as we add a proton or neutron to build up heavy nuclei. The gap in the middle indicates the location of ^{5}Li, which is not stable. This makes it hard to create ^{6}Li, which is stable, or any other heavy nuclei.

This process explains why very few elements heavier than helium are produced in the early universe. It also explains why roughly 90 percent of the atoms in today's universe are hydrogen, with helium making up almost all the rest. More importantly, this clarifies why the abundance of hydrogen and helium is the same in every direction: both were created everywhere in the thermal equilibrium of the hot, early universe.

We note, for now, that it will take about half a billion years before heavy atoms like carbon and oxygen (needed to produce life on Earth) form. This formation will not take place in outer space, but in the hot, dense center of stars (Chapter 16). When a star dies and explodes, it jettisons its atoms into outer space.

Before we leave the three-minute mark, it is worth commenting on the particles in the universe. We note that electrons and positrons have mostly annihilated each other; hence, there are almost no positrons left. However, for reasons we still do not fully understand, there were slightly more electrons than positrons in the early universe. Since it appears there was no net charge when the universe was created, we are left with one electron for every proton. It is primarily the fundamental building blocks of nature—like protons, electrons, and photons—that move about in space.

Since neutrons can only live outside a nucleus for about fifteen minutes, it will not be long before all the neutrons in the universe have either been incorporated into nuclei or have decayed away (see Figure 14.3). This affects the varying number of nuclei types in the universe. From here, it will take a long time for further drastic changes to occur.

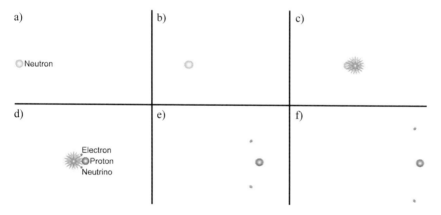

FIGURE 14.3 A neutron decaying into a proton, electron, and neutrino. If a neutron is not part of a nucleus, it will typically decay fifteen minutes after being created.

14.2 A FEW HUNDRED THOUSAND YEARS AFTER THE BANG: ATOMS FORM

Around the universe's three hundred thousandth birthday, it is big enough and cool enough that photons are not always knocking electrons out of atoms. Since any atom created now is more likely to stay intact, over the next couple hundred thousand years, the universe fills with stable atoms. This process essentially allows all the electrons, protons, and light nuclei in the universe to combine into atoms as they are highly attracted to each other because of their electric charge.

This period of time—when all the atoms combine and stay combined—is called **recombination**. Frankly, I would have called it "combination," but the name is historical and has stuck (see Figure 14.1). The time right in the middle of recombination is currently thought to be 380,000 years after the bang and have a temperature of about 3,000 Kelvin.

Figure 14.4 shows three stages in the transition from a hot to a cool universe. It starts as a high-density place made up chiefly of charged particles, then transitions into a low-density phase, still occupied primarily by charged particles. In the third stage, the universe has expanded and, after recombination, is filled with neutral particles.

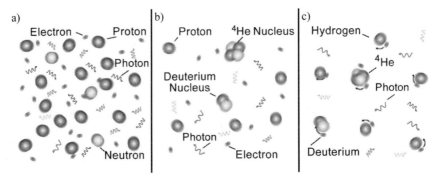

FIGURE 14.4 As the universe cools, it transitions from a high-density place of mostly charged particles; through a phase of low density with mostly charged particles (many of which are deuterium and helium); and, eventually, into a universe with neutral atoms after recombination.

The bathtub analogy described in the last chapter for atoms is shown in Figure 14.5. In the early universe, there are very few atoms because they have been effectively broken apart by photons. Later on, we have lots of atoms because all the free nuclei and electrons have combined and can now stay that way.

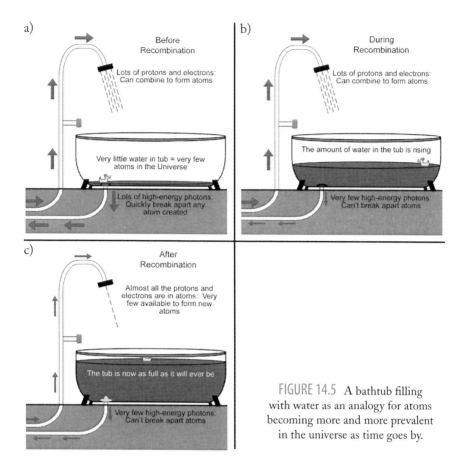

a) **Before Recombination**

Lots of protons and electrons: Can combine to form atoms

Very little water in tub = very few atoms in the Universe

Lots of high-energy photons: Quickly break apart any atom created

b) **During Recombination**

Lots of protons and electrons: Can combine to form atoms

The amount of water in the tub is rising

Very few high-energy photons: Can't break apart atoms

c) **After Recombination**

Almost all the protons and electrons are in atoms: Very few available to form new atoms

The tub is now as full as it will ever be

Very few high-energy photons: Can't break apart atoms

FIGURE 14.5 A bathtub filling with water as an analogy for atoms becoming more and more prevalent in the universe as time goes by.

After the nuclei and electrons have combined, the universe becomes "transparent" for photons. Since all protons and electrons are in atoms, they largely ignore photons unless the photons have a high-enough energy to break the atom apart, or have the special energy required to excite the atom (see the laws of quantum mechanics in Chapter 7, in particular Figure 7.7).

Before recombination, photons interacted—and scattered—with the protons and electrons they met (see Figure 14.6). After recombination, the universe resembles the bottom part of the figure. The ability of photons to now travel uninterrupted is one of the reasons we can see galaxies from so far away. It is also why photons in the cosmic background radiation look the way they do: they stopped interacting with atoms when they reached a temperature of about 3,000 Kelvin. Because of the universe's expansion, these photons are now about 2.7 Kelvin.

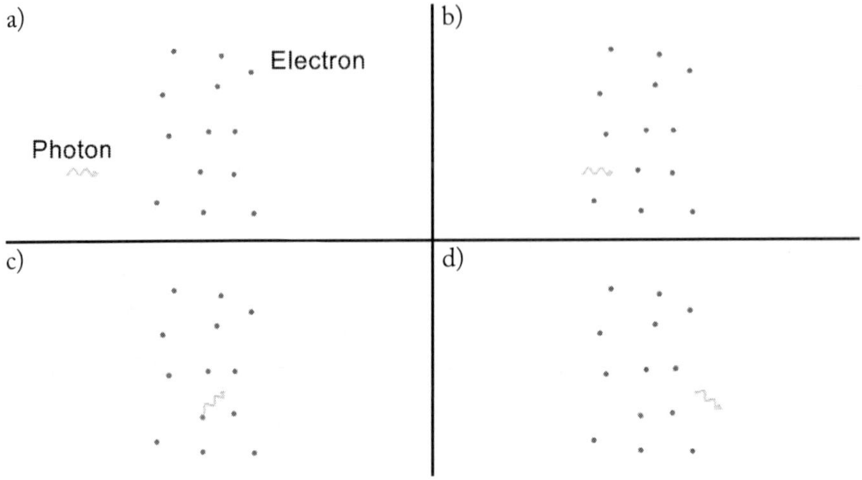

FIGURE 14.6 A set of action shots showing a photon interacting with matter before and after recombination. In the top set of frames, the photon interacts with electrons and changes direction. Since our eye would be able to see the light—but only after it scattered from the charged particles—the universe looks more like frosted glass than like window glass. In the bottom set of frames, after recombination, most photons travel in straight lines forever and do not interact with atoms very much because of quantum mechanics. The universe now looks transparent.

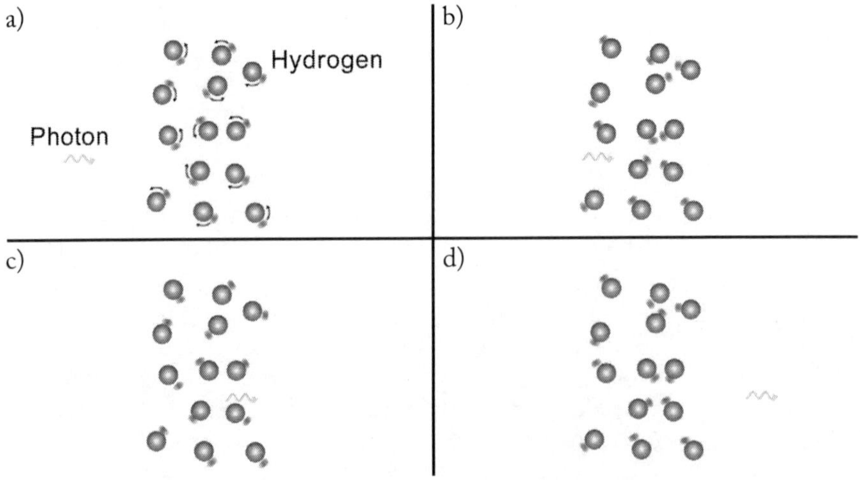

An analogy is useful here. Today, photons travel straight through space like they travel through clear glass. Protons and electrons in the universe *before* recombination prohibited photons from traveling in a straight line, scattering the light as if it were shining through heavily frosted glass. This is what makes it effectively impossible to see further back in time (farther away) than a few hundred thousand years after the bang. Then again, there are no objects like stars or galaxies that far away because they would not yet have been created.

14.3 THE NEXT FOURTEEN BILLION YEARS

After recombination, the universe is dominated by the interactions of neutral particles. Since the charged particles (electrons and protons) have combined into atoms, essentially all the particles in the universe are now neutral. Thus, there is a change in the dominant way the particles interact with each other. Between a second after the beginning until approximately the three hundred thousandth year mark, the charged particles have mostly interacted by means of electromagnetism (attraction and repulsion of charged particles, as described in Chapter 7) and photons breaking things apart.

In many ways, photons dominated the universe in much the same way that a bully dominates a schoolyard—by relentlessly breaking things apart. But after recombination is over, photons no longer dominate. Now that the overwhelming majority of particles in the universe are neutral, the dominant way particles interact is via gravity. This has two effects. First, the universe will not change quickly because gravity is such a weak force. Second, since there is no longer anything that produces a repulsion force, all the things in the universe are attracted to each other. A major exception to this rule applies to dark energy, which we will discuss in Chapter 18.

The gravitational attraction between individual particles draws them together. The reason that all atoms are not quickly drawn to a single spot in space is that the universe expanded an enormous amount over the first couple hundred thousand years. By this time, the universe is already sparsely populated.

Luckily for us, as beings with lots of combined atoms, there was an adequate amount of individual atoms in close proximity to each other. As two atoms get closer together, they create more mass in the place they occupy. Since, between them, they generate a bigger dent in space-time, they can then attract even more mass. This progression continues, and the dent keeps expanding.

About a half a billion years after the bang, there are a number of places around the universe where enough mass has gathered that galaxies and, eventually, stars can form. This process will be described in Chapter 15. We note for now that dark matter will play an important role.

The first stars appear as lots of atoms get close together. Each atom picks up speed as it falls to the center of the star because of the attraction of gravity. Ultimately, there are enough high-energy atoms in the center of the star that the conditions become hot and dense like they were a few minutes after the bang. This allows hydrogen atoms to fuse (like in Figures 8.3) to create helium and photons.

There is a difference, however, between this situation and the early universe that is important to note: this time, the atoms are all held near the center of the star by gravity, so the number of helium atoms in a star rises quickly. Thus, it is just a matter of time before there are enough helium atoms in a small space that heavy elements like carbon, oxygen, iron, and other materials can be produced. If a star explodes—and many will—it will throw these atoms into the universe. There, the atoms can amass into planets and, eventually, into human beings.

While planets can be created around any star, our Sun and Earth will form at about the nine-billion-year mark. Fast forward fourteen billion years after the big bang to the present day, and we have planets (at least one) that have aged and produced life. This process will be described further in Chapter 16.

We have finished our description of the very early universe and how we got from there to the universe of today. It is now time to discuss in more detail how stars, galaxies, and black holes are formed.

Massive Things: Galaxies, Stars, and Black Holes

In the last unit, we finished our quick history of the universe. This unit features some of the most massive things in the cosmos, as well as their formation and their roles in this remarkable history. The chapters are:

Chapter 15: Galaxies
Chapter 16: Stars
Chapter 17: Black Holes

We will see that gravity dominates the ways galaxies and stars form in the evolving universe and discuss how both began appearing a few hundred million years after the bang. For that reason, we could easily have reversed these next two chapters. We also note at this point that even though we did not discuss when dark matter came into the universe in the previous unit, our best understanding—for reasons we will discuss in Chapter 19—is that it all came into existence a tiny fraction of a second after the universe began. Thus, it is an integral part of galaxy formation.

We will also describe stars second because they can "die" in interesting ways, depending on the individual star and its properties. In some cases, they can form black holes, which are among the oddest things in the universe. Black holes deserve a chapter unto themselves, and they get one at the end of this unit.

Galaxies

We dedicate this chapter to an in-depth look at galaxies—some of our universe's biggest and most impressive constructions. While scientists still have a lot to learn, we can talk with some confidence about what galaxies look like, and how and when they formed during the evolution of the universe.

We begin with an overview of two types of galaxies, known as **spiral galaxies** and **elliptical galaxies**, as shown in Figure 15.1. We then describe the first billion years of the universe, with an eye to how the first galaxies began to form. While star formation is an important part of this story, and will be mentioned briefly, star births will be discussed in more detail in the next chapter.

NASA/JPL–Caltech/ESA/Harvard-Smithsonian CfA

NASA, ESA, and the Hubble Heritage (STScI/AURA)-ESA/Hubble Collaboration

FIGURE 15.1 The top two pictures show different types of galaxies. In the top is a spiral galaxy, like our Milky Way. In the center of the galaxy is a bulge of stars; some bulges are like miniature versions of an elliptical galaxy, which is shown in the middle figure. Outside the bulge is the disk of a spiral galaxy in which stars orbit around the center together. Within the bulge, or an elliptical galaxy, stars orbit the galaxy in many directions, as shown in the bottom.

Stars not to scale
Not all stars shown

Star

Dark Matter Halo

Center of Galaxy

After we have talked about what galaxies look like, we will turn to how they form as well as some of the evidence for this knowledge. We then conclude with our best understanding of when galaxies formed in the history of the universe.

15.1 What Galaxies Look Like

While the universe hosts many different types of galaxies, all of them have important properties in common: they contain mostly stars (which produce light), gas (atoms not in stars), and dark matter. Most of the mass is in dark matter, but most of the light we see comes from stars. Many of the atoms are dispersed throughout the galaxy and are not in stars.

To visualize how the dark matter is spread around the galaxy, you can think of the galaxy as a giant ball of dark matter as seen in Figure 15.1. If you think of a ball with water in it, like a fish bowl, then the dark matter completely engulfs the stars and atoms the same way the water surrounds the fish. This isn't a perfect analogy because water has the same density everywhere, whereas there is more dark matter near the center of the galaxy and there is dark matter far beyond the outer stars in a galaxy. Scientists typically refer to the mass that surrounds the stars as the **dark matter halo**.

The stars are easier to visualize in a galaxy as they provide most of the light. While the stars look very densely packed in a galaxy, this is merely an illusion. Instead, what we're usually seeing are billions of stars many light-years apart. There are, of course, exceptions to this rule, including binary stars—two stars orbiting each other.

Developmental properties are what make each galaxy look different. For the sake of brevity, we'll pass on the opportunity to discuss all the different types of galaxies, but instead, quickly describe two types and point out important features of each. After we describe them, we will travel back to their beginnings, coming full circle to their shape today.

The first type we will consider is the one we live in: a spiral galaxy.

A spiral galaxy possesses some interesting structure as we saw in Figure 2.3. For example, the densest part of the Milky Way galaxy is in its center and the stars there form what is known as the **bulge**. Outside the bulge is what we call the **disk**. The stars in the disk far outside the center of the bulge orbit in an organized manner, in some ways orbiting our Sun like planets.[1] Each star in the bulge also orbits around the galaxy's center, but less like a horse on a merry-go-round and more like a bee buzzing around its hive (with the density of bees being closest near the center).

The second type of galaxy we consider is an elliptical galaxy, which doesn't have many distinguishing features. However, even though they do not have a disk, elliptical galaxies can be much larger than spiral galaxies; the biggest galaxies are elliptical galaxies. The stars orbit the center of the galaxy in a fashion that is similar to the way stars in a spiral galaxy orbit in the bulge.

15.2 A GRAVITY-DOMINATED UNIVERSE AND GALAXY FORMATION

While the seeds of galaxies first appeared in the very early universe (more on this soon), it is gravity that causes the matter in the universe—mostly dark matter—to form what we now call galaxies. By the few-hundred-thousand-year mark, when the universe was cool enough for electrons and nuclei to remain together as atoms, most of the charged particles in the universe had combined to become neutral particles. Electromagnetic forces between particles were therefore almost completely eliminated (except inside or very close to an atom). With gravity unimpeded, atoms and dark matter could come together over the next half a billion years to form the first galaxies.

To understand the creation of galaxies, we consider two analogies. One is how mass collects in space and the other is how it forms the different types of galaxies themselves.

We visualize the initial collecting of large amounts of matter to form a galaxy by thinking of little kids jumping on a trampoline (Figure 15.2). Each child represents some mass (like an atom) and the trampoline represents space-time. As the kids jump, they create dents in the trampoline. If two kids collide, they fall into the trampoline and create an even bigger dent in space-time, making it more likely that other children will fall into their dent. As more kids fall, each dent gets bigger until everyone who was nearby falls into one of the giant dents.

---

[1] Understanding how the "spiral arms" formed in the disk is more complicated and beyond the scope of this book. For more information on them and how they are created, see the texts in the Suggested Reading.

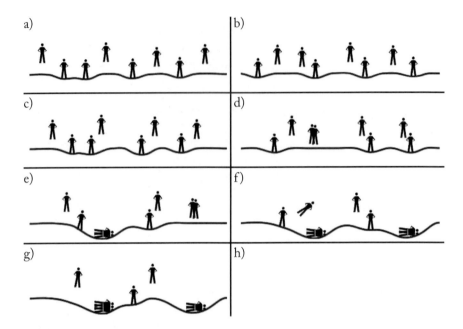

FIGURE 15.2 An analogy of galaxies forming around the universe using kids jumping on a trampoline. Slowly, as the kids collide, they create bigger dents in the trampoline (space-time). It is these enlarging dents that will become the beginnings of the billions of galaxies around the universe.

It is the same way with matter. Essentially, once you get a big dent in space-time, all the nearby matter starts falling into it. These big dents are where galaxies form.

Visualizing how matter collects in a big dent in space-time will require some creative thought and a number of imperfect analogies. Each helps visualize some of the next stages of galaxy creation, as well as some of the reasons why we have different types of galaxies.

Since humans don't have a lot of experience with matter moving toward the center of a large dent in space-time, we will describe different examples we are used to in daily life and then come back to galaxies. One example is water being poured into a bowl and flowing to the bottom. Another example is water swirling in a toilet bowl. A third is water in a bathtub with the drain open (and ignoring what happens to the water after it goes down the drain). If you have ever watched water go down a drain, you have seen that it can do so in a number of different ways, depending on the size of the drain and the amount of water in the tub. A big dent in space-time is like a large bowl. If there isn't much water in the bowl, all the water will fall straight to the bottom and collect there. However, if the bowl is full, the water farther away from the drain will bump into the water in the middle. Sometimes this produces swirling.

Basically, three things are important: (1) how deep the dent in space-time is, (2) whether the matter in the dent in space-time interacts very much, and (3) how much matter there is. Going back to the previous examples, at the atomic level (which we cannot see with our naked eyes), as water moves in a bowl, the water molecules are bumping into other water molecules in front of them, slowing them down—much like movie patrons jostling each other as they exit a crowded theater. This explains one of the reasons we see a whirlpool after a toilet is flushed; some of the water quickly goes down the drain, but the water on the outer part bumps into the water in the middle and goes off in a different direction. Toilet bowls are designed so that the outer water swirls before going down the drain. If we look carefully, we can see that the water closer to the center is moving more quickly than the water near the outer edge of the swirling—much like a coin in a gravity well (for more on why this occurs, see Box 15.1).

BOX 15.1

Some people prefer a different analogy for atoms speeding up as they both move around the center of the galaxy and fall toward the middle. Imagine a spinning ice skater twirling with her arms out during the Winter Olympics. Her arms are like the atoms, and her body is like the center of the galaxy. When she brings her arms in, she spins faster and her arms move around her more quickly. You can try this at home in a rotating chair. Get yourself spinning with your arms and legs stretched out, and then pull them in. Your rotation speeds up when you bring your limbs in. This is what physicists call **conservation of angular momentum**.

We now turn back to galaxies and a time before lots of atoms and dark matter have converged into a galaxy. Both atoms and dark matter move within a forming galaxy, but it's important to remember that each interacts differently. The atoms are like the people jostling to get out of the crowded movie theater. If there are lots of them and they all move toward the door at the same time, from different directions, they will bump into each other and slow down. However, if they all move together, they can quickly get out of the theater. Similarly, atoms being pulled to the center of the galaxy will bump into each other and fall toward the center or can swirl around the center of the galaxy forever, like swirling water in a perfect toilet bowl where the water never slows down or goes down the drain.

Dark matter, on the other hand, does not interact much, so it can flow almost freely.

With all this in mind, we pick up our story again about half a billion years after the universe started. At this point, the universe is still very dense everywhere, but gravitational attraction among particles has collected groups of dark matter and atoms together and formed billions of clumps around the universe where the dent in space-time is a little deeper than average. You can think of the collection of matter as a giant cloud, millions of light-years across. As time goes by, each cloud gets smaller due to the pull of gravity toward its center. We also note that these clouds are often bumping into each other and merging.

As the cloud contracts, different parts of it will fall to the center in slightly different ways. The densest part of the galaxy will be at the center of the galaxy (like the bulge in our Milky Way), and some of the matter will clump together and move around the center in a stable orbit that can go on forever—like the Sun orbiting the center of the Milky Way.

15.3 STARS IN GALAXIES

Scientists are still working out the details of how different types of galaxies form. For now, we will focus on a simplified explanation of how stars form and move within a galaxy and say more about stellar formation in the next chapter.

As each atom falls toward the center of a galaxy, its proximity to other atoms greatly increases. With enough atoms in a small area, atoms can interact and begin forming stars. While only a portion of the atoms in a galaxy condense into stars, as time goes by, more and more stars develop and they continue orbiting the galaxy's center. Since a star is an incredibly massive object, most of the atoms anywhere near it will either fall into it or orbit around it. Once a star "turns on," the light emitted from the star will push against the nearby atoms, planets, and other debris that may have formed. In the case of our own star, the Sun, the planets, asteroids, and other heavy things are too massive to be pushed out of the Solar System, but many of the individual atoms have long since been cleared away.

While it is not always true (for example, in binary stars), stellar formation often makes the average distance between stars very large—so large, in fact, that many stars are so far apart they do not collide very often. At this point, they stop acting like water going down a drain, and act more like a planet going around the Sun; in principle, they can keep orbiting the center of the galaxy forever.

Switching to the formation of stars in the disk in spiral galaxies, we focus on the atoms farthest away from the center of the galaxy. This process is shown in Figure 15.3. In the early days of the galaxy, they feel only a small gravitational pull toward the center. These distant atoms can thus move slowly around the center of their galaxy for hundreds of millions of years.

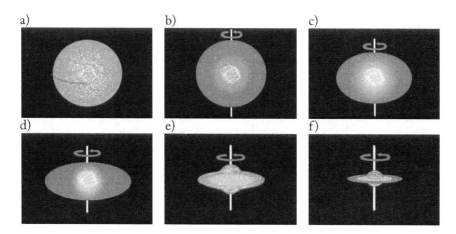

FIGURE 15.3 A simplified illustration of how atoms in a spiral galaxy, like the Milky Way, form into a disk-like shape. Here we have ignored collisions between galaxies as they form. In some sense, the galaxy is "spinning into shape." Note the bulge in the center and the flattened disk around it. This shape makes the spiral galaxy look similar to planets going around the Sun.

With the passage of time, the galaxy contracts under the pull of gravity and the atoms gradually move closer together and start to interact more. Again, because these are atoms, they interact like people trying to leave a crowded movie theater. As this happens, the atoms that follow along with the crowd are able to easily move; the smooth-traveling path is an orbit around the center since they can move in lock-step. However, atoms that try to move in a different path are much more likely to get jostled by the main crowd and bumped. Thus, it is just a matter of time before most of the atoms that were moving "the wrong way" either start moving with the crowd or fall toward the center of the galaxy, leaving only those atoms that are in-step with the crowd.

Over time, as only atoms moving in the special orbit continue (and ignoring any matter that falls in from a collision with a neighboring galaxy), the outer parts of the galaxy assume a "flattened" appearance. Eventually, the only atoms that continue orbiting are the ones that move in a single plane—what scientists call the **equatorial plane**.

Ultimately, as these atoms move together over millions of years, they will be close enough to be attracted to each other, bump into each other, and form stars. Eventually, the outer part of the galaxy will look like a flat, rotating disk of stars and gas. This is the disk part of a spiral galaxy.

In the next chapter, we will see other examples of this flattening and of this collecting of atoms into large objects when we talk about planets forming in a plane around an equator.

The bulge of a spiral galaxy, and elliptical galaxies which resemble them, are more complicated and can form in a number of ways. One of the common ways they form is through collisions between galaxies. When galaxies collide, many of the stars are knocked out of their orbits, but the two galaxies merge because of their mutual gravitational attraction. The stars eventually orbit the center of the combined galaxy, and can look like bees buzzing around a hive.

Since the Milky Way is on a collision course with the Andromeda galaxy, there is good reason to believe that both galaxies are likely to get ripped apart, and then combine again under their mutual gravitational attraction to form a much bigger elliptical galaxy.

15.4 WHEN DO GALAXIES FORM?

Having completed our simplified description of how galaxies form and what they look like, we next turn to when they form and their earliest beginnings.

The most distant galaxies we have observed sent us their light from about half a billion years after the bang. Measurements of the ages of the stars provide a separate estimate. The oldest observed stars are measured to be about 13.2 billion years old, consistent with being produced half a billion years after the bang.

Understanding when galaxies start to form in the history of the universe turns out to be quite complicated. Simple calculations that predict how long it takes for galaxies to form in a universe that evolved through perfect thermal equilibrium suggest it would take much longer than the current age of the universe for the first galaxy to form. What has gone wrong? What do we need to add? What else needs to be taken into account?

Surprisingly, the answer comes from the role of quantum mechanics in the early universe. Because of quantum mechanics, there can be small variations in the amount of mass at different places in space—in an atom, for example, where most of the mass is in the nucleus, but there is basically no mass between the nucleus and the first electron in its orbit (see Figure 3.2). It turns out that in the earliest moments in its history, when the universe was smaller than an atom, the laws of quantum mechanics had a big impact on its density in any given place. Scientists call these **quantum fluctuations**.

To understand why small variations in the amount of matter in space can be important, consider that a place with a slightly greater amount of matter is more likely to be the place where matter clumps together over time. This means that galaxies form at that site before they form elsewhere.

Let us go back to our trampoline analogy for more detail.

If we start with all the kids equally spaced, then it will typically take awhile for the first two to collide. If the two started out slightly nearer each other, this area will be slightly denser and they are, on average, more likely to collide sooner, and therefore create the first large dent sooner. Similarly, fluctuations in the density of the universe would impact how long it takes to create that first big dent in space-time; the earlier it happens, the quicker galaxies show up in time.

Is there any evidence for these density variations in the early universe? Figure 15.4 shows the temperature map of the universe originally seen in Chapter 12. The color represents the temperature at each location in the sky. It appears to be the same everywhere, with a temperature of 2.728 Kelvin. But is it perfectly uniform? The answer is no. If we show the *difference* between the temperature and the average temperature (by subtracting the same temperature from all places), we can begin to clearly see small differences.

These temperature disparities, displayed in the top right part of the figure, show that it is slightly hotter in the upper right corner of the map (where it is red) and an equal amount cooler in the lower left corner (where it is blue). One might think this means that the universe does not have the same temperature everywhere. However, in Chapter 9, we described that this is what it looks like if you are moving inside a bunch of atoms that possess the same temperature: it appears hotter on one side than on the other, just as in the figure.

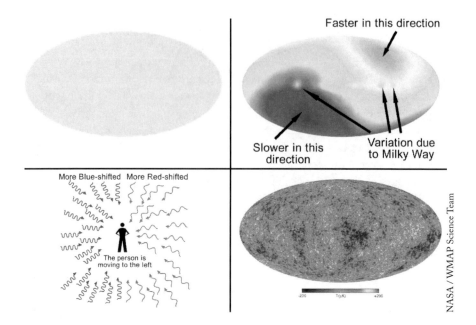

FIGURE 15.4 The temperature of the universe. The top left shows the temperature distribution of the universe in terms of color in all directions. The top right shows the temperature distribution after we subtract the same amount in every direction. The big variations (color differences) are due to Doppler shifts because we are moving relative to the cosmic background radiation; the little variations in the middle are from the Milky Way's light. We can subtract off both of these effects, as well. The results are shown on the bottom right. These small temperature fluctuations in the cosmic background radiation give us confidence that our model of when galaxies form is correct. We will come back to these temperature fluctuations in later chapters. (Color versions on page C-9.)

As the Solar System orbits around the Milky Way, and the Milky Way moves through space, all motion must be taken into account as a Doppler shift. The photons that are moving away from us are red-shifted (lower temperature) and those moving toward us are blue-shifted (higher temperature). This makes it look hotter in one direction and colder in the other. Taking the Doppler effect into account (again subtracting it off), with a speed of several hundred kilometers per second, we can see what other effects are left.

The temperature variations, after subtracting the Doppler effect (see the bottom right of Figure 15.4), are about 200 microKelvin, or 0.0002 Kelvin. Comparing this to the average temperature of 2.728 Kelvin is like measuring the difference between 70.000 degrees Fahrenheit and 70.005 degrees Fahrenheit. Your skin cannot tell the difference between those two temperatures, but very sensitive instruments can.

To understand how small this is, consider stacks of pennies, each $1,000-worth high. This is about 500 feet tall (almost two football fields). The variation is like adding or taking away a single penny from the top of a stack here or there; an extra penny added makes the red spots in the figure, and a single penny taken away makes the blue spots.

The variations are consistent with the expectations of quantum fluctuations in the early universe, as well as the pull of gravity in galaxy formation. The areas of slightly higher density form the earliest seeds of galaxies. Using the measured densities, scientists predict that the first galaxies formed a few hundred million years after the bang. This was just before the ages of the oldest galaxies we observe (meaning the light we see emitted billions of years ago). This remarkable story hangs together fairly nicely.

Since the earliest stars appeared around the same time as the earliest galaxies, we next turn to star formation. Along the way, we will discuss star deaths in Chapter 16 and black holes in Chapter 17. We will come back to quantum fluctuations in the early universe in Chapter 20 when we talk about inflation.

Stars

The pull of gravity makes for some similarities between the way stars and galaxies form. In this chapter, we focus on the birth of stars, as well as their life and death. We will also talk about black holes and how they fit into the picture. In the next chapter, we describe different types of black holes, some of the evidence for them, and some of their properties.

The way a star lives and dies depends mostly on its mass. Whether or not there are other stars nearby also plays a role, since stars can interact with each other in lots of complicated ways. We will focus on isolated stars—those hanging out alone—with our Sun as a prime example. We will also discuss a few other types of stars and other massive objects like **black holes, red giants, white dwarfs, supernovas, and neutron stars** (see Table 16.1). We start with the birth of the first stars a few hundred million years after the big bang.

16.1 ATOMS IN STARS

The first galaxies formed after a few hundred million years due to the influence of gravity pulling together lots of mass. Galaxies start off as regions millions of light-years across, and it is in these regions that smaller collections of atoms come together to form stars. One can ask, "Which came first: galaxies or stars?" This is kind of like a chicken-and-the-egg question. Most scientists agree that galaxies contain stars, so they might say that stars came first. Then again, others say that since stars can only form in galaxies, galaxies came first.

Focusing on a set of atoms that are near each other in a galaxy, gravity lures them to a nearby place containing a lot of mass (the largest nearby dent in space-time). As was explained in our simple description of spiral galaxy formation in the last chapter, a resulting dense clump of matter will gradually form in the center of this same area, since it is now rich in atoms. If there are an adequate number of atoms in this region, the clump of matter will eventually become a star.

As atoms fall into the star, they pick up speed, become very energetic, heat up the center of a star, and start to interact, further increasing the temperature. Other atoms orbit around the star; these can end up in planets, asteroids, and comets (see Figure 16.1). We will describe the atoms in the star and the atoms that orbit the star separately.

Buzz word	What is it?
Star	A typical star is a giant ball of atoms in which individual nuclei are continuously fusing in the core to form heavier nuclei. This process creates light that we can see here on Earth.
Galaxy	A giant collection of massive objects, typically containing billions of stars and even more dark matter
Red giant	A stage in the life of many stars, including our Sun. Our Sun will live about ten billion years (it is about four and a half billion years old now). It will eventually become a red giant for about a billion years before becoming a white dwarf. During the red giant phase, most of the helium will reside in the star's small "inner core," while the hydrogen will reside in a huge outer shell that emits mostly red light (thus the name "red giant").
White dwarf	For stars like our Sun, after the red giant stage, they shed their outer shell and only their hot inner core remains. The remnant is known as a white dwarf and typically has about the same mass as our Sun, but the same size as the Earth.
Supernova	The explosion that occurs when, for example, a really massive star runs out of fuel. In this case, the core collapses under the force of gravity, protons and electrons combine to form neutrons, and the energy released causes a giant explosion. What remains at the core is typically a neutron star or a black hole. A star can also explode as a supernova if it gains mass from eating another star (see Chapter 18).
Neutron star	An extremely dense object that can remain after a supernova. In this case, it has collapsed into an incredibly dense ball of neutrons. It has the same density as if the mass of the Sun was condensed into a small city (~10 km) or the mass of the Earth were contained within a marble.
Black hole	What remains after a star has been crushed down to a single point in space under its own gravitational weight. It is so incredibly dense, even light cannot escape, which is why it is called black.

TABLE 16.1 "Buzz words" in astronomy.

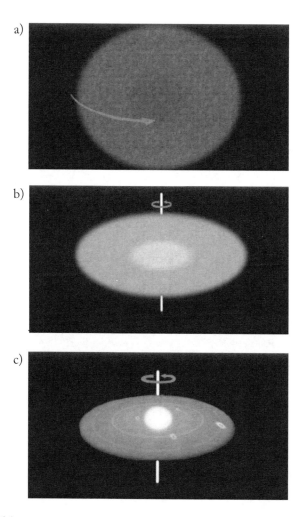

FIGURE 16.1 Stars form in a small region of space where lots of hydrogen and helium atoms have collected. Like in the last chapter's simplified description of the creation of a spiral galaxy, this collection of atoms begins to spin and the rotating cloud contracts and flattens. Eventually, since most of the atoms are in the center, this becomes the star. Planets, like the Earth, can grow from the gas and the dust in the disk that surrounds the star.

Since the inside of a star contains large numbers of high-energy atoms in a small place, the atoms can interact in much the same way that particles did in the early universe. For example, many of the atoms in a star will get broken apart into nuclei and electrons (see Figure 12.5). From here on out, we will focus on nuclei and the way they interact with each other.

In the same way as demonstrated in the figures in Chapter 8, when energetic nuclei get close together, they can fuse and produce energy as well as the light that we can see. If a star has enough atoms to start with, then it only takes a few million years before large quantities of nuclei will be continuously fusing. This process brings about the birth of a star, which can shine for billions of years.

A decent, but perhaps oversimplified, definition of the word "star" is that it is a "collection of nuclei held together by gravity that continuously fuse together to create heavier nuclei, and in the process, create light." Then again, there are other types of stars (like a white dwarf) that don't fit this definition. For the sake of simplicity, we will refer to an object as a star if it actually was a star at some point in its lifetime.

Before continuing, we will quickly mention atoms outside a star. As with our simple description of the formation of a spiral galaxy (see Figure 15.3), atoms that orbit a star can eventually form a thin disk. Within this disk, those atoms near each other can clump and form planets like the ones in our Solar System.

Mirroring this process on a lesser scale, material rotating around planets can slowly form smaller objects, such as rings or moons. We note that the formation of planets like our Earth would not happen for billions of years after the big bang because the first generation of stars will initially need to create heavy atoms like oxygen, carbon, nitrogen, and iron in the universe. We also note that the creation of our Moon and the rings of Saturn are much more complicated than the process described above.[1]

Getting back to stars, we will use metaphors to describe them as if they were people: they have a birth, a life, and a death. We can also describe stars like cars: they run on "fuel" and they "die" when they run out of fuel. They also are like balloons—giant collections of atoms "forced" to stay inside surrounding walls.

Now that we've discussed how stars are born, we'll move on to looking at the fuel in a star and the process that keeps atoms together like a balloon during the star's lifetime.

16.2 NUCLEAR REACTIONS AND GRAVITY KEEP STARS ALIVE AND MAKE THEM SHINE

Why don't all the atoms that are not orbiting simply fall to the center of the star? On Earth, things fall toward its center, but they stop because they hit the Earth's

---

[1] For more information on both, see the books in the Suggested Reading.

surface. Do stars like our Sun have a surface? Does the answer have anything to do with the interactions among the atoms that produce light? To answer these questions, we must look in more detail at how stars operate. As we will see, the atoms give a star its "life," which is effectively a battle between gravity trying to crush the atoms into a tiny point and the interactions among the atoms that "oppose" the force of gravity.

We start our description by returning to the simple balloon analogy (see Figure 16.2). Hydrogen and helium are the gases inside the balloon, and gravity is like the Mylar™ or latex that keeps the atoms inside. While this is a useful analogy, stars are admittedly more complex than simple balloons. Gravity can push harder than a typical balloon shell for one thing. For another, the nuclear reactions inside stars give the atoms the energy they need to firmly press against the shell of the balloon, keeping it inflated. Gravity, of course, is not an actual shell, but rather a force that keeps the atoms together. Nuclear reactions only occur at much higher temperatures than those found here on Earth. This is why fusion reactors are hard to make work effectively to create the power needed to run a city.

To understand further why gravity acts like a shell, consider a single atom inside a star as it is pulled to the center by gravity. The bottom parts of Figure 16.2 show a high-powered rifle, as well as an experimental gun, that can fire bullets at different speeds near massive objects like the Earth.

In the first configuration, a person on Earth shoots a bullet into the air using a rifle. Just as the bullet falls back to the Earth, an atom with a low speed (low energy) is pulled back toward the center of a star. It cannot leave the star because of the pull of gravity.

Similarly, the rifle cannot shoot a bullet fast enough to leave the Moon because of the Moon's large mass. However, the rifle can make a bullet travel fast enough to leave an asteroid—an object with a much smaller mass. In addition, experimental guns have been built that can shoot a bullet fast enough to leave the Moon.[2] The analogy here holds as well: if an atom has a high speed, it can leave the star.

Ultimately, then, both the speed and the mass of the object matter. If the atom has a low speed or if the star's gravity is strong (as it is in stars), gravity drags most of the atoms back down again so they stay inside the star. While gravity is not a true wall, the more massive the star, the stronger the pull and the harder it is for atoms to escape.

---

[2] We can now understand why people build rockets to leave the Earth. Like a car engine, rockets use their propulsion systems to keep speeding up while they travel. A rifle, however, gets only one chance to fire a bullet with a high-enough speed to exit the Earth.

FIGURE 16.2 While the Mylar™ part of a Mylar balloon (picture on top right) keeps its helium gas from escaping with a thin wall, gravity holds hydrogen and helium atoms inside a star. Lots of nuclear reactions among the hydrogen (and helium) atoms are what push against the gravity. In our analogy, the atoms push against the walls of the balloon and keep it inflated. The nuclear reactions also produce the light we see. In the bottom two rows, we show four different configurations that are analogous to an atom in a star with gravity. If the atom is low-energy (low-speed), or if the star is very massive, the gravity will keep the atom inside the star. If the atom has a high energy (large speed), or the star is light, then the atom can leave.

What gives atoms the high speeds they need to avoid being crushed by gravity into a single point at the star's center? To understand the answer, we need to come back to the interactions among these atoms.

Since the interactions in a star are essentially between nuclei, we refer to these as **nuclear reactions**. Many of these interactions are identical to those present in the early universe when nuclei were initially created (see Chapters 8, 12, and 14). For example, we can have a collision between two protons (hydrogen), that combine to form deuterium (^2H); a collision between a deuterium nucleus and a proton, that combine to form helium (^{3}He); or a collision between two ^{3}He nuclei, that combine to form ^{4}He and two protons. The protons formed through the latter interaction are available for future nuclear reactions.

The crucial thing about all these interactions is that the new nucleus is always lighter than the two lighter nuclei used to create it. For example, deuterium is lighter than two protons. What makes this important is that in the reaction, the extra mass is converted into other types of energy (remember $E=mc^2$), and a small amount of mass is a huge amount of energy. This energy typically manifests itself in the form of very high speeds (energy) for the atoms, or high-energy photons that come directly from the nuclear reaction.

These high-energy atoms and photons will then interact with the other atoms and photons throughout the star, gradually producing large numbers of low-energy photons, as discussed in Chapters 7 and 9. Some of these photons make their way out of the star and travel across the universe to us. These photons are what we see here on Earth and are what make us say that stars "shine."

The high speeds of the nuclei produced in the nuclear reactions also keep the star from collapsing. Each reaction speeds up the atoms and "presses" against gravity. In our analogy, they "push" to make the balloon bigger. Thus, what we have can be thought of as a balancing of two forces: the energy of the nuclear reactions pushing to make the balloon bigger, and the force of gravity trying to make the balloon smaller.

It is just a short amount of time after the birth of a star that these two forces balance each other out and the star reaches a size at which it is stable. The size of the star depends on the number of atoms inside it—in other words, the mass of the star. Figure 16.2 demonstrates this balance. Once the star has reached this balancing point, it can remain stable for billions of years as it uses up its fuel (hydrogen and helium), all the while creating the light that we see.

16.3 A Star is Born

We next focus on the birth and early life of stars. Before a star starts shining, you can think of it as a "ball" made up primarily of hydrogen. Stars of different masses lead different lives, and throughout its life, the size of a star changes as the atoms inside it fuse.

For simplicity, we start by describing stars like our Sun, which are situated far apart from any neighboring stars in space. As long as the mass of a particular star is similar to the mass of the Sun (we say our Sun has one **solar mass**), it will behave in much the same way. In the beginning, atoms fall toward the center of the star, making this location where the density of the star is the highest. We call this area the **core** of a star. See Figure 16.3. As atoms fall in, they speed up and heat the core. Nuclear reactions between the atoms will also heat the core and its temperature will inevitably rise above the 10 million Kelvin mark that is needed for hydrogen nuclei to routinely combine and fuse into helium. We call this the birth of the star. The core temperature of our Sun is now about 15 million Kelvin.

We note quickly that fusion in the core only happens regularly for stars with a mass of more than roughly 8 percent or more of the Sun's mass. Lower than about 8 percent of the mass of the Sun, the temperature will never get high enough to fuse hydrogen; since it doesn't have fusion that produces light, in some sense, it isn't really even a star. We refer to such an object as a **brown dwarf.**

In a star like our Sun, fusion regularly converts hydrogen into helium in the core because it is very hot and dense there. The atoms outside the core rarely fuse

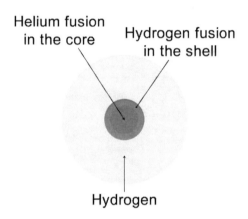

FIGURE 16.3 A star starts as a ball of mostly hydrogen and eventually forms a core in which hydrogen fuses to create helium. It is here that the energy is released that powers the star and ultimately produces the light we see. At a later stage in the life of a star, shown in the bottom figure, the hydrogen in the core has been used up and a star fuses helium in the core. A shell of hydrogen forms outside the core and fuses hydrogen. In all cases, the rest of the star is mostly hydrogen outside the shells and core, and does not fuse much.

because they possess a lower temperature and a lower density. It is important to note that the core of our Sun is not yet hot enough to fuse helium nuclei; that requires a temperature of about a hundred million Kelvin. Thus, the helium "ash" will simply collect in the core of the star.

Hydrogen fusion is what makes our Sun produce photons (makes it shine), keeps it stable, and slowly converts its hydrogen into helium. As long as there is adequate hydrogen in the core, interactions will take place in which nuclei fuse and release energy. When the star gets older, the core begins to run out of hydrogen (run out of fuel) and its ability to produce light and high-energy atoms is diminished. Without the nuclear reactions to "push" against it, gravity wins the battle and compresses the atoms closer and closer together. What happens next depends on the mass of the star, since variations between them can be huge.

In general, stars possessing a mass less than eight times that of the Sun (eight solar masses) will go through a **red giant** stage and eventually become a **white dwarf**. More massive stars typically go through additional stages, during which they build up heavier elements. These gigantic stars can eventually explode as a **supernova**, end up as a **neutron star**, or collapse directly into a **black hole**. While we will describe all of these in turn, we next focus on stars like our Sun.

16.4 LIFE AND DEATH OF STARS LIKE OUR SUN

For all mid-sized stars like our Sun (possessing between roughly 8 percent and eight times the Sun's mass), life will follow the same basic path, in that they will all end up as a white dwarf. However, the details are different for the different masses. For this reason, we will just take our Sun as an example; its future is shown in Figure 16.4. Our Sun will burn the hydrogen in its core for about ten billion years; it is about 4.5 billion years old now. When the Sun exhausts the bulk of the hydrogen fuel in its center, what remains in the core is mostly helium. Since the temperature will not be high enough for helium to regularly fuse, the rate of nuclear reactions will significantly diminish, marking a time when the star enters a new stage in its life.

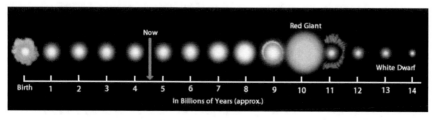

FIGURE 16.4   The life cycle of our Sun, which is a typical star of medium mass. The Sun is about 4.5 billion years old. (Color version on page C-10.)

Without enough nuclear reactions to "oppose" the crush of gravity, the core of the Sun will compress. We will focus on the Sun's core, but note that this time begins the red giant phase of the star's life as the outer part of the Sun will expand, for reasons discussed a bit more in Box 16.1. The name "red giant" comes from a description of the outer part of the star that can be seen from large distances.

BOX 16.1

While the red giant phase of a star is quite complicated, a simplified understanding can be had. The red giant phase begins when most of the hydrogen in the core of a star is used up. At this point, since there are not enough nuclear reactions to support the core at its full size, the core (now mostly helium) contracts under its own weight. During this process, the helium atoms fall in, speed up, gain energy, and heat the core of the star to higher temperatures. At the same time the hydrogen atoms surrounding the core then fall toward the center, pick up energy, come closer to each other, and form a shell around the helium core. With a higher density and energy, these hydrogen atoms now can also start the fusion process and produce a large amount of energy that further heats the core as well as the atoms in the outer part of the star. This pushes the outer parts of the star to expand ten times farther out into space; in some cases, it can be a hundred-fold. For our Sun, it is expected to be a considerable fraction of the distance from the Sun to the Earth (which would be bad for the Earth). This giant outer part is cooler than the star that preceded it, thus it glows light that is of lower energy, so it looks redder. This is where the name "red giant" comes from. The red giant phase typically lasts between half a billion and a billion years.

We also note that at the end of the life of the helium-burning star, like our Sun will be near the end of its life, there will be shells of helium and hydrogen around a carbon core. The nuclei in these shells will fuse and heat up the outer part of the star which causes it to expand. This time, the expansion is so large that the outer part of the star can be blown away from the core of the star. What remains is our white dwarf.

In high-mass stars (more than eight times the mass of the Sun), a core can use up its hydrogen, then the new core can use up its helium, and the process repeats with a core of heavy nuclei and multiple shells with successively lighter nuclear build up around them as the star ages. You can think of these shells as layers like an onion. Near the end of its life we can have a hydrogen shell outside a helium shell, outside shells of carbon, oxygen, neon, magnesium, silicon, and ultimately iron ash in the core.

As the star's core contracts, atoms are pushed toward its center. Eventually, the atoms are so densely packed that the other atoms in the core start to push back. As they get close to each other, there is a particular repulsion among the electrons in the atoms that balances out the attraction of gravity. Basically, it is a quantum mechanics thing. In much the same way that electrons can only orbit at certain distances from the proton (described in Chapter 8), electrons can only exist near each other at certain distances.[3] When a balance is reached between the push of electrons and the attraction of gravity, the core stops contracting.

The atoms outside the core, mostly hydrogen, can now fall toward the center and create a shell around the helium core. We call this a **hydrogen shell**. As they fall in, they pick up speed, heat up, start fusing, and heat the helium core to even higher temperatures. They also cause the outer part of the star to heat up and expand, which is why the star becomes a giant. It can stay this way for about a billion years.

Once the core's temperature reaches about a hundred million degrees, enough energetic helium has been crammed into a small enough space that the helium atoms can start fusing into heavy nuclei. The process of **helium burning** has begun. The center of the star is now a helium burning core, with an outer hydrogen burning shell as shown in Figure 16.3. While stars with a mass of more than twice as much as the Sun are more complicated (and can create oxygen), we will stay with our description of stars like our Sun.

Helium burning has a couple of very quick steps. In the first step, because the helium nuclei have very high energy and the core is very dense, two helium nuclei can regularly find each other and combine into beryllium (Be)—^{8}Be, to be more precise (see Figure 16.5). Back when the star was young, any beryllium created would decay back into two helium nuclei almost as quickly as it was produced; it has a lifetime of 2.6×10^{-16} seconds (which is a tiny fraction of a trillionth of a second). But at this stage in the life of a star, helium burning produces so many beryllium atoms in such a small space that there is a good chance that a newly formed beryllium atom will meet a helium atom before it decays. When this happens, they can combine to form carbon (^{12}C), as shown in the bottom of Figure 16.5.

This process is important not only because it releases more energy—which can show up as light—but also because ^{12}C is stable. This allows it to build up as carbon "ash" in the core. In the early universe, heavy nuclei could not build up by merely adding one proton or neutron at a time, as shown in Figure 14.2. But now,

---

[3] This is known as the **Pauli exclusion principle**, which states that identical particles, like electrons, cannot get too close to each other. Scientists also use the phrase **degeneracy pressure** to describe the push electrons give against gravity. For more on these topics, see the books listed in the Suggested Reading.

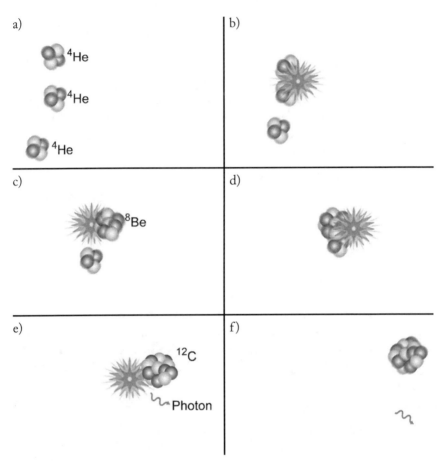

FIGURE 16.5 In a later stage of life, if a star with a helium core gets hot and dense enough that the helium nuclei can fuse to form beryllium (which only lasts for 10^{-16} seconds), but that is long enough to find another helium nuclei and form carbon, as shown in the bottom figures.

heavier elements in the universe *can* build up. It is these heavy stable nuclei, created in stars billions of years ago, that eventually formed our planets and, in time, us.

It is just a small matter of time (about one hundred million years) before the helium in the star's core is used up and the core is basically left with carbon. Since the mass of the Sun is too small to have enough atoms to create temperatures high enough to fuse carbon (about six hundred million Kelvin), the nuclear reactions stop and the core shrinks once again. While the outer part of the star gets blown away (again see Box 16.1), just the carbon core remains and this is what is known as a **white dwarf**. Stars with a mass heavier than the Sun, but less than eight times heavier, will have both carbon and oxygen in the core.

While a white dwarf is not fusing atoms, since it was essentially just the hot core of a star, it remains white-hot. Since we see it shining white light, this is part of where it gets its name. Since it was just the core of a star, a white dwarf is also incredibly dense; while it can be 60 percent the mass of the original star, it can also be the size of the Earth (our Sun is about a hundred times wider than the Earth). This is why it is called a dwarf. A white dwarf is so dense that a pair of standard-sized dice made from white dwarf material would weigh about five tons! Our Sun will become a white dwarf star in about five billion years.

Over the ensuing millennia, the Sun, like most white dwarfs, is expected to eventually convert all its energy to photons and other forms of radiation, stop shining, and the remaining atoms will likely become what scientists have dubbed a **black dwarf**. However, it is believed that this process takes so long that, as of yet, no stars anywhere in the universe have evolved into a black dwarf. Indeed, there is no evidence of a black dwarf ever having been observed, and the coolest observed white dwarfs give an estimate of the limit of the age of the universe.

16.5 LIFE AND DEATH OF MASSIVE STARS

The life of stars much more massive than the Sun can be similar to their smaller mass counterparts for awhile, but there are important differences. This includes how long they live, as well as what happens at the end of their life.

We start by noting that the larger the mass a star starts with, the sooner it will die. Even though they have more fuel to begin with, the crush of gravity due to their larger mass crushes them to be denser in the core. The denser the star, the faster nuclear reactions occur (and the brighter it shines), so the sooner it runs out of fuel. Part of the reason it stays stable is because the more massive a star, the hotter it needs to be to balance out gravitational attraction.

While a medium-sized star like our Sun can live for about ten billion years, stars that are ten times more massive than the Sun only last thirty million years. The most massive stars—a hundred times more massive than the Sun—barely last a hundred thousand years. This is one of the reasons why there are so many more small mass stars in the universe: most of the most massive ones have already burned out.

We next turn to the lives of high-mass stars. They start out similarly to stars like our Sun, in that they will also go through a hydrogen burning and helium burning phase. However, while the mass of the Sun is not able to create the temperatures needed for carbon fusion, larger mass stars can produce those high temperatures.

The details of the life of a high-mass star are complicated, but throughout its life it can create carbon and oxygen, which can further combine to make even heavier elements that will eventually produce life here on Earth. In most cases, when these heavy nuclei form, more nuclear energy is released. Sometimes, really heavy nuclei are formed and they can decay to release energy. However, as shown in experiments here on Earth, once iron is created (either from fusion or decay) there is no more energy you can get from it. Iron "ash" builds up in the core and once everything lighter than iron is used up in the core, the fuel is gone and we are left with a lot of iron atoms. After the fuel runs out, the nuclear interactions are not able to push back against the crush of gravity. From there, things happen very quickly.

Again, let us consider separately what is at the center and in the outer part of the star.

In the star's core, gravity crushes the atoms into a space so small that electrons are pushed so close to the protons that the two interact. All at once, the electrons and protons combine and convert into neutrons, and neutrinos that fly off into space. In many ways, this is like neutron decay described in Chapter 13, only turned around; this process will be described more in Chapter 19.

After all the protons and electrons in the core combine, we are left with a densely packed ball of neutrons with a diameter about the size of Manhattan. This is known as a **neutron star**, even though it is not a star in the ways in which we are familiar. A neutron star is often referred to as a **stellar remnant,** as it is what remains after the core of a star has used up its fuel.

The conversion from the iron core to a neutron star is the beginning of a giant explosion called a **supernova**. It is not too hard to understand why. The gravitational energy from the neutrons falling to the center, along with the energy released into the neutrinos, is enormous.

A second important feature of this supernova is that the atoms falling from outside the core, and hitting the core, are like a rubber ball dropped from the top of a tall building onto a hard sidewalk. As it hits the sidewalk, it can get squished before it bounces back into the air. As the atoms outside the core fall toward the center, they reach very high energies and there are lots of heavy atoms in a small space. In the area outside the neutron core, with all these neutrinos and high-energy heavy atoms, we will get lots of nuclear interactions which will create very heavy nuclei, like gold and uranium.

Between the neutrinos leaving and the atoms bouncing back into space, the giant explosion carries the remaining atoms (but not the neutrons) into space. The speeds and violence of this explosion are enormous; a wave of material rushes into space with speeds up to 30,000 km/s (a tenth the speed of light). The stuff that became our Earth arose from this type of explosion billions of years ago.

These explosions can be so violent that the light produced can outshine a small galaxy for several weeks or months. During this short time, a star can expel most, or all, of its energy into space. It can also radiate as much energy as the Sun emits over its entire lifespan.

I like comparing a supernova to a dandelion. At the end of its life, a dandelion—like a giant star—turns from yellow to white, eventually goes "poof!" and sends its seeds out to create the next generation.

On average, supernovae occur about once every one hundred years in a galaxy with as many stars as are in the Milky Way. Scientists can observe about one a day somewhere in the universe.

There are a number of other types of supernovae, one of which we will come back to in Chapter 18. Astronomers are not worried about them causing damage to the Earth since there are no good candidates nearby.

If a neutron star is not too massive—less than twice the mass of the Sun—then it can remain a neutron star for quite a while as it uses up the rest of its energy. What keeps it from getting crushed is the fact that neutrons, like electrons, do not like to be "too close" to each other. The repulsion of the neutrons balances out the squeeze of gravity; everything can balance and the star can stabilize again. At this point, everything is tightly compressed; the star is not unlike a giant neutron-only nucleus. Neutron star material has a density of billions of tons per cubic inch, which means that a single marble made from a neutron star material would weigh about the same as our Earth!

If a neutron star has a mass greater than three times that of the Sun—what we call a "critical mass"—it is just a matter of time before gravity overcomes the repulsion between neutrons. In this case, there is nothing left to oppose the crush of gravity. The star collapses under its own weight until it becomes a **black hole**. Indeed, if the star is heavy enough, it can develop directly into a black hole after it becomes a supernova.

We will talk more about black holes in the next chapter.

Black Holes

Movies often portray a black hole as a demonic, power-sucking vortex. However, this is not how scientists understand them. As described in the last chapter, a black hole is just another thing a star can turn into when it runs out of fuel. In this case, it is simply a massive ex-star—or stellar remnant, to be more precise—that has completely collapsed under its own weight. Then again, any object that has the mass of a star, but has all its mass in a space smaller than a proton, is going to have some unusual properties.

In this chapter, we will describe more about why we call them black holes and other ways to make black holes. We will also discuss what a black hole would look like if you were close to it, evidence for black holes, one of the reasons why black holes may be important in our understanding of galaxies and, perhaps, the big bang itself.[1]

17.1 WHAT IS A BLACK HOLE?

To better understand a black hole, we focus on the impact of large amounts of mass on space-time.

We start by considering our Sun (see Figure 17.1). Currently, our Sun is more than one million kilometers across (more than a hundred times bigger than the Earth) and puts a big enough dent in space-time to keep all the planets in orbit. But if the Sun were somehow shrunken down to only a few kilometers across (it is not clear how this could happen, but we can think about it hypothetically, as is done in Box 17.1), then it would have the same mass, but the higher density, of a neutron star. Thus, it would make a much deeper dent in space-time.

---

[1] There are a lot of fun other topics, like microscopic black holes, worm holes, and Hawking radiation, that are incredibly exciting and could have been included in this chapter. We will not cover them here as there is no evidence for them yet; you can learn more about them in the Suggested Reading.

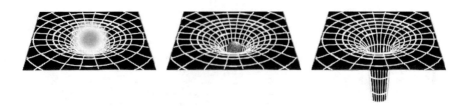

FIGURE 17.1 An artist's impression of various dents in space-time. The figure on the left is the dent created by our Sun. The middle figure shows the deeper dent made by a neutron star. The right figure shows what the dent looks like for a black hole. The depth of the dent is infinitely long, so we have cut it off for this illustration.

BOX 17.1

What If Our Sun Were Magically Compressed So Much That It Became a Black Hole?

Scientists often think about hypothetical situations like a sudden—and magical—transformation of the size of Sun to be smaller than a proton. In this case, the Sun's mass would not change, just its size. Inside the Sun, the curvature of space-time would be very large. However, the gravitational force on the planets (far outside the Sun) would not change, so they would continue to orbit as they always have. Eight planets orbiting "nothing" might look odd to an outside observer, but that is what they would see. From our vantage point on Earth, it would take eight minutes for the last light emitted from the Sun to reach the Earth, so it would be eight minutes before we knew the Sun had shrunk. After that, we would not see any light from it at all. Without light from the Sun, it would get cold on the Earth quickly and we would all die.

This resulting dent would be deepest inside the star, while the outside of the star would not change much. In fact, at a distance far away from the Sun—say at the location of the Earth—there would not be much difference in the pull of gravity at all. This would be true no matter how small the size of the Sun shrinks. If the Sun were somehow even further compressed, into a black hole, its density would become so immense that the curvature of space-time at the black hole's center would be effectively infinite. You can think of this as an endlessly deep hole; what was a dent in space-time has become more like a cliff.

But why is a black hole black?

We return here to the analogy of shooting a bullet into the air. The bullet will come back down because of the pull of gravity. Even if the bullet is shot with a really fast speed, it will still return to Earth since even the best guns are not that powerful. How fast does it need to be shot so that it leaves the gravitational pull of the Earth? Physicists figured this out centuries ago and the knowledge helped rocket scientists build spacecraft which have left the Earth, landed on the Moon, and even gone beyond many of the planets.[2] The speed needed to leave the Earth is called its **escape velocity.**

The escape velocity for other objects, like the Sun, depends on the mass and size of that object. It is smaller for the Earth than for the Sun (see Figure 17.2). A bullet must be shot at more than 11 km/sec to leave the Earth. (As previously noted, this is why rocket scientists use rockets—they can keep speeding up as they travel.)

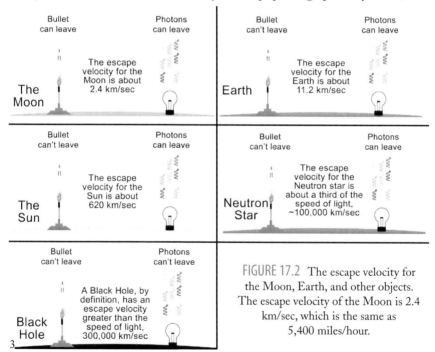

Bullet can leave	Photons can leave
The Moon The escape velocity for the Moon is about 2.4 km/sec	

Bullet can't leave	Photons can leave
Earth The escape velocity for the Earth is about 11.2 km/sec	

Bullet can't leave	Photons can leave
The Sun The escape velocity for the Sun is about 620 km/sec	

Bullet can't leave	Photons can leave
Neutron Star The escape velocity for the Neutron star is about a third of the speed of light, ~100,000 km/sec	

Bullet can't leave	Photons can't leave
Black Hole A Black Hole, by definition, has an escape velocity greater than the speed of light, 300,000 km/sec	

FIGURE 17.2 The escape velocity for the Moon, Earth, and other objects. The escape velocity of the Moon is 2.4 km/sec, which is the same as 5,400 miles/hour.

Since the Moon is smaller, a bullet only needs to be going about 2.4 km/sec to be able to leave. For the Sun, a speed of 620 km/sec is needed. This means that anything traveling faster than 620 km/sec can escape from the dent that the Sun puts in space-time, while anything moving more slowly will be pulled back.

[2] Voyager 1 and Voyager 2, each launched in 1977, are now more than seventeen billion kilometers from the Sun, beyond the orbits of Neptune and Pluto.

The escape velocity for a typical neutron star is just over a third the speed of light (~100,000 km/sec), so only things that are moving faster than that speed can escape.

If our Sun were shrunken to the size of a neutron star, its escape velocity would also be about a third the speed of light. If it were shrunken even further, the escape velocity would get correspondingly bigger. At some point, it becomes so dense that the escape velocity is faster than the speed of light (300,000 km/sec). Since nothing we know of can go that fast, even light cannot escape. A black hole, loosely speaking, is an object with an escape velocity greater than the speed of light.

The implication is remarkable: light does not have a high enough speed to leave a black hole. Doesn't light always travel at the speed of light? If it cannot slow down, shouldn't it be able to leave? The answer to the first question is "yes," and the answer to the second is "no."

The path of light in curved space-time is bent in three dimensions (see Chapter 6). One way to imagine this is that the "space" part of space-time is "falling" into the black hole.

Think of yourself walking on an escalator or a moving walkway like you see in an airport (Figure 17.3). If you walk up the escalator at a certain speed, but the escalator moves downward at a greater speed, the net effect is that you go downward. In the same way, if the walkway moves backward faster than the speed of light, then light cannot move forward along this path—even though it moves at the speed of light.

The bottom line is that if light cannot escape, we cannot see any light coming from the star, so it appears black.

Before moving on, we quickly describe two different types of black holes: **stellar** and **supermassive**.

Stellar black holes are the result of a supernova or a neutron star that has collapsed. They typically have masses a few times that of our Sun. Supermassive black holes, on the other hand, are much more massive than stellar black holes. It is not known exactly how they formed, but they have been found at the centers of many galaxies and can possess masses ranging between hundreds of thousands of times to billions of times that of our Sun.

There is a supermassive black hole located at the center of our Milky Way galaxy, and there is some evidence that there may be one at the center of all large galaxies. It is very possible that they play an important role in galaxy formation and

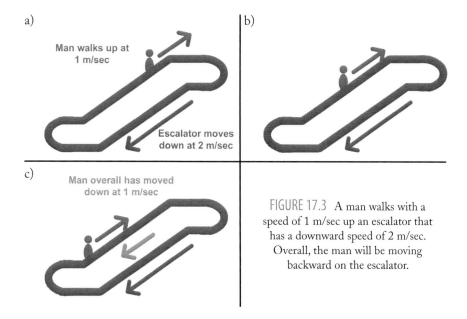

a)

Man walks up at 1 m/sec

Escalator moves down at 2 m/sec

b)

c)

Man overall has moved down at 1 m/sec

FIGURE 17.3 A man walks with a speed of 1 m/sec up an escalator that has a downward speed of 2 m/sec. Overall, the man will be moving backward on the escalator.

evolution. Scientists are not certain of this, however, since they do not clearly understand how supermassive black holes form. However, they have observed that the larger the galaxy, the bigger the supermassive black hole in the center. Perhaps this is an important clue.

17.2 Life Near a Black Hole and Evidence for Black Holes

With no light coming from a black hole, it is hard to discover and study them, and perhaps impossible to "see" what is going on inside. As of this writing, only about twenty stellar black holes have been observed near us, with the closest being about three thousand light years away (see Figure 2.3d). Since it is interesting to think about what these neighbors look like close up, let us pretend to send out an astronaut or a robot for a more detailed look. This will help us understand how black holes were discovered and how they are studied today.

Shining a flashlight on a black hole is useless because when flashlights shoot photons at an object, our eyes only see the photons after they bounce off the object we are trying to view. In this case, a photon in the beam of light aimed at a black hole would fall into it and would never come back out. This is one of the reasons we use the word "hole" to describe them; other ways to study them are needed.

Anything that gets too close to a black hole will fall in and never come back out. The ability to indirectly study black holes can be enhanced, then, by understanding how these things fall into them. To do so, we return to our analogy of objects near the surface of the Earth.

We said that a bullet needs a speed of 11 km/sec to escape from the surface of the Earth. However, if the bullet started in the upper atmosphere, it would only need a smaller speed (less than 11 km/sec) to escape since it is farther away from the Earth's center. The more distant from the center, the smaller the velocity needed to escape.

The **event horizon** for a black hole is the specific distance from its center where the velocity needed to escape is exactly the speed of light. Anything that gets closer to the black hole than the event horizon cannot leave; it will never again be seen by anyone outside the black hole. The more massive the black hole, the farther out the event horizon is from the center. A black hole with a mass of five times that of the Sun would have an event horizon of 15 kilometers.

The event horizon can help us indirectly study black holes as we watch things approach them. It is important to realize that space-time is so stretched that a person on the outside of a black hole will see things very differently from a person falling in. Let us consider sending an astronaut toward a black hole (see Figure 17.4) from both perspectives.

We start with the astronaut's perspective. As he falls in, bad things happen to him. Since the space-time curvature is changing so quickly, the gravitational pull on the side of the astronaut closest to the black hole would be much stronger than on the side farther away. Our unlucky space traveler would thus get stretched out and then ripped apart, a process given the suggestive name of **spaghettification**. We could not hear the screams over the radio because radio transmissions are radio waves, which are low-energy photons; like all photons, they cannot escape.

Turning to our perspective, we consider what would happen if we gave the astronaut a flashlight to send a signal for each second that ticks by as it moves toward a black hole, but before it passes the event horizon. As the flashlight nears the event horizon, the space-time around the black hole gets stretched, so the signals appear to get farther and farther apart. Additionally, the wavelengths would become stretched, which means they would be red-shifted. Thus, he and the light would look "redder and redder." Eventually, as he gets infinitely close to the event horizon, the wavelength would get infinitely stretched and we wouldn't be able to see it any more.

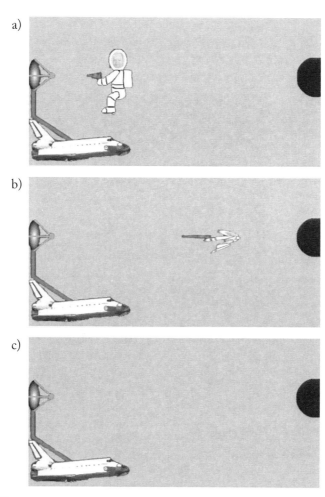

FIGURE 17.4 An astronaut shooting a photon gun near a black hole. The antenna attached to the spaceship can observe the photons from the gun until the astronaut nears the event horizon. As he gets closer, the wavelengths would get stretched (get redder and redder) and eventually, he would completely disappear from view. From the perspective of the astronaut, since the attraction of gravity is stronger near the center of the black hole than farther away from it, the astronaut would get stretched out and ripped apart as he gets closer and closer.

Scientists have used these ideas to hunt for black holes in space. If two stars are orbiting one another, and the heavier one turns into a stellar black hole, the two can continue to orbit each other. What we see, though, is a star orbiting around "nothing."

In some cases, the gravitational attraction of the black hole can be large enough to rip off the outer atoms from its neighbor (especially when it enters its red giant phase). A good analogy for these atoms as they fall into the black hole is that of crumbs on the surface of water as it goes down a drain. The crumbs swirl faster and faster around the drain as they approach it. See Figure 17.5.

X-ray: NASA/CXC; Optical: Digitized Sky Survey

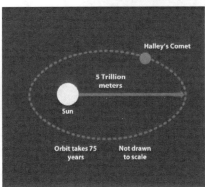

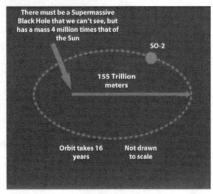

FIGURE 17.5 In the top row, we see a stellar black hole (left side of the drawing), which can be identified as it rips off atoms from a nearby star (right side of the drawing). As the material falls into the black hole, it produces the two beams of light we can see. (Color version on page C-16.) On the bottom row, we show examples of how we can use what we know about gravity to measure the masses of stars and black holes. By watching Halley's Comet orbit around the Sun (how long it takes to go around and how far it travels from the Sun), we can measure the mass of the Sun. Similarly, from the orbit of a star that orbits a black hole, we can measure the mass of the black hole. Supermassive black holes can be discovered and measured in similar ways. A star (SO-2 in the figure) orbiting around "nothing" near the center of the Milky Way provides evidence that there is a supermassive black hole there that we cannot see directly. The supermassive black hole in the center of the Milky Way is more than four million times the mass of the Sun.

In the same manner, matter is pulled inward toward the black hole and moves increasingly fast in a giant swirl. As this happens, the atoms are forced closer together, collide, and produce light (mostly X-rays) that we can see here on Earth. By observing the X-rays (from material near the event horizon) and the neighboring star (far from the event horizon), we can determine that a star is, indeed, orbiting a black hole.

Scientists can also measure the mass of a black hole by studying the orbit of a nearby star. To calculate the mass of the Sun, for instance, scientists can measure the distance of Halley's Comet from the Sun and the amount of time it takes for the comet to orbit it (as shown in Figure 17.5). In the same way, measurements of stars as they move around a black hole can be used to measure the black hole's mass.

A supermassive black hole at the center of the Milky Way galaxy was discovered and measured in this manner. A number of stars near the center of the galaxy were found to be orbiting around "nothing." By measuring the path of each star as it moved around the supermassive black hole (how quickly it orbited and how far it traveled from the black hole in its orbit), the mass of the supermassive black hole was measured. This black hole at the center of our galaxy is now known to be some four million times the mass of the Sun.

17.3 WHAT BLACK HOLES MAY TEACH US ABOUT QUANTUM MECHANICS, GENERAL RELATIVITY, AND THE BIG BANG

In the 1970s, scientists Stephen Hawking and Roger Penrose realized that the big bang and black holes may have a lot in common. If we are to understand a black hole as having all of its mass at a single point in space—what scientists call a **singularity**—then this is what the universe might have been like at its very beginning.

Hawking and Penrose speculated that the creation of a black hole, as it collapses from a star down to a tiny point in space, is like the big bang in some ways, but running backward in time. Said differently, the creation of a black hole is some mass in a finite amount of space being crushed into an infinitesimally small amount of space, while the big bang would be some amount of mass in an infinitesimally small amount of space that quickly expands into a finite (and eventually very large) amount of space. If this were true, then by studying one, you are studying the other, and perhaps learning more about both at the same time.

This is an exciting possibility, but the details are a problem. General relativity describes how things operate in curved space-time, but does not work well when space-time is infinitely curved.

Another problem is that at the time of the big bang, space should have been really small—smaller than an atom, at some point—and the laws of quantum mechanics should be important at these sizes. That is not a problem in itself, but our understanding of quantum mechanics does not properly describe curved space-time. In other words, quantum mechanics and general relativity are inconsistent with each other. Their predictions contradict each other for calculations inside a black hole and we do not know which one (if either) is correct.

Ideally, we want a single theory to describe nature. Like good detectives, scientists have tried to put these two theories together to see if they tell a consistent story. Despite our best attempts, though, there is currently no good "quantum theory of gravity" (also called a "theory of quantum gravity") that can tie the two together into a single "unified" theory. It may turn out that something like **string theory** is the answer. Speculations are rife about this topic, but that discussion is for another book (see the Suggested Readings).

In the meantime, we will have to stay tuned. Perhaps by studying black holes, we will finally figure out what is going on. With a "more correct" understanding of the laws of physics, perhaps we can determine what happened at the big bang itself, or if the universe began with a bang at all.

We have now concluded our unit on galaxies, stars, and black holes. Next we will move to Unit 6: Early Times, Dark Energy, and the Fate of the Universe.

Early Times, Dark Energy, and the Fate of the Universe

The previous two units outlined the evolution of our universe from an age of about a millionth of a second to the cosmos we observe today. This last unit offers our best prediction as to the ultimate fate of the universe, as well as our best understanding of times earlier than a millionth of a second. We do not know as much as we would like about either of these, but scientists are learning more all the time.

There are three chapters in this unit. They are:

> Chapter 18: Possible Fates of the Universe, Dark Matter, and Dark Energy
> Chapter 19: Particle Physics, Dark Matter, and the Very Early Universe
> Chapter 20: Inflation and the Earliest Moments in Time

Chapter 18 covers some of the potential fates of the universe and the data available to help us predict which one is in store for us. It also discusses dark matter and dark energy, as well as their role in both the universe's evolution and its ultimate fate.

Chapter 19 goes into more detail about the fundamental particles in nature and their interactions with each other. Since the very early universe contains these particles interacting at very high energies—and high-energy collisions can change one type of particle into another—we need a full account of which particles can exist in nature to fully describe this period in our history. At the same time, we will explore the possibility that dark matter is a fundamental particle, which could further illuminate the connection between particle physics and cosmology.

Finally, we conclude in Chapter 20 with the idea known as "inflation," which describes the earliest moments of the universe; times such as 10^{-40} seconds after the bang.

Keep in mind that while the data are powerful, many of the ideas in the following chapters are not fully proven and that pieces of the puzzle are still missing. New data and theoretical advances have recently changed our understanding of the universe. The case is far from closed.

Possible Fates of the Universe, Dark Matter, and Dark Energy

While the ultimate fate of our universe is unknown, we can use what we know to make predictions. Will it continue expanding forever? If not, could gravity slow the expansion of space-time so much that—like a thrown ball falls back to the Earth—the cosmos would slow down, come to a stop, and start collapsing back in on itself?

The answer depends on a number of things, such as how fast the universe is currently expanding, how much material is in the universe, how big it is now, and whether we understand all the laws of physics. As in previous chapters, we will start with the simplest understanding, as if all the laws of physics are already understood. As we will quickly recognize, however, this last assumption is not at all accurate. We will therefore also discuss what we *do* know when it comes to our modern understanding of the universe's future—a discussion that includes the role of dark energy.

This look at the universe is very similar to our first depiction of its expansion in Chapter 10. We begin by gaining a basic understanding of how scientists think about its fate. When we are better prepared, we will move to a more thorough exploration that includes the latest data.

18.1 THREE SIMPLE MODELS FOR AN EXPANDING UNIVERSE

If the expanding universe is only filled with known particles and the expansion is solely governed by known forces, there exist but a few possible fates. Since gravity continually tries to pull everything in the universe back together, then each of these possibilities depends on three important factors: the amount of mass in the universe, the size of the universe, and the present speed of expansion.

We can predict that if the universe is expanding fast enough, then it will expand forever. Similarly, there is the possibility that gravity will eventually overcome the current expansion and crush the universe back down again. A third possibility for the ultimate fate lies exactly between these two possibilities.

To understand why the mass, size, and speed of expansion are important, think back to the bullet analogies in Chapters 16 and 17 (Figures 16.2 and 17.2). One way to ask whether a bullet will be able to leave an object, like the Earth or an asteroid, is to consider its speed relative to the escape velocity; it is either above, below, or exactly equal to that velocity.

Another way is to focus on the object that the bullet is leaving instead of the bullet itself. If the bullet has a speed slightly more than an asteroid's escape velocity, it could leave the asteroid's gravitational pull. The same speed bullet on the Earth, however, would not be able to escape.

Let's apply this same analogy to a hypothetical mystery star. If a bullet had a certain speed—say 100 km/sec—at the center of the star, we can ask: What would the mass of the mystery star need to be so that the bullet's speed is fast enough to escape? We call this the **critical mass** or, taking into account the size of the star, its **critical density**.

We are now ready to consider our three cases.

The first possibility is that the universe will keep expanding forever. This is like shooting the bullet into the air with a speed that is higher than the escape velocity. In this case, the bullet will keep flying "away" forever. Said differently, if the density is smaller than the critical density, the gravitational pull will not be able to overcome the speed of expansion of the universe. This possibility is shown in the middle of Figure 18.1 with the universe expanding forever.

The second possibility is that the mass in the universe will slow down the expansion so much that, in time, it will stop expanding, start contracting, and eventually end in what we call a **big crunch** or **Gnab Gib** (which is big bang backwards). This outcome is like a bullet with a speed below the escape velocity, or an object with a density bigger than the critical density. In this case, the bullet will rise into the air, but eventually, it must crash back down to the surface. If the speed of expansion of the universe is small, or if the density of the universe is large, then space-time will eventually stop expanding and start the contracting process. It has been speculated that our universe could contract into a single black hole. This is illustrated on the left side of Figure 18.1.

a)

b)

c)

d)

e)

f)

g)

h)

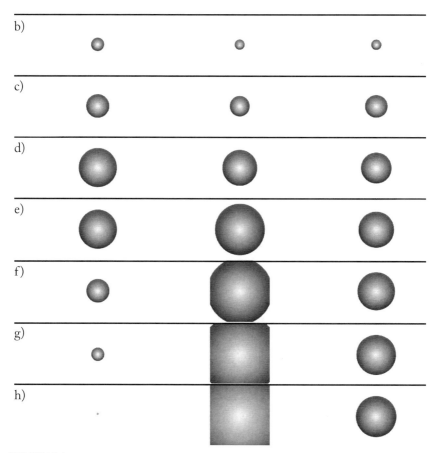

FIGURE 18.1 Three series of action shots using our two-dimensional analogy illustrating the possibilities for an expanding universe (read top to bottom). Left column: A universe that expands for awhile, then contracts back down to nothing (big crunch). Middle column: A universe that expands forever. Right column: A universe that expands for awhile, but slowly stops growing and will stay around forever. It is important to remember that the universe is just the surface of these simple pictures.

The third possibility is a delicate balance between the other two cases: the universe expands forever, but with a rate slowly decreasing toward zero speed. In our analogy, the bullet is shot exactly at escape velocity. In other words, the universe possesses precisely the critical density. This possibility is shown on the right side of Figure 18.1. It is an exact balance between the universe's expansion rate and density.

A simple illustration for these three scenarios showing the size of the observable universe over the course of its history is shown in Figure 18.2. Note that we don't know the true size of the universe, and what we can talk about with confidence is actually the visible universe. If you like, you can think of how the distance between far-separated galaxies is changing over time. The challenge for us as detectives is to determine which, if any, of these scenarios corresponds to the universe in which we live.

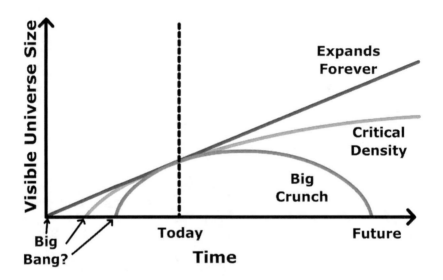

FIGURE 18.2 A graph showing three simple possibilities for how the size of the visible universe could change over time. The graph assumes that all the mass and energy in the universe can be directly seen, and that dark energy is not taken into account. On the vertical axis, we see the size of the visible universe; on the horizontal axis, we see time. In all three cases, we see space expanding and getting bigger. In the big crunch scenario, the universe gets bigger for awhile, and then goes back down to zero size. The other scenarios will expand forever. In this simple description, the fate of our universe depends on its density and on how quickly it is expanding.

To figure this out, we need to know how fast the universe is expanding today and its density. We can measure the speed of expansion of the universe fairly accurately (see Chapter 10). This uniquely determines the value of the universe's critical density. Perhaps if we could obtain an accurate measurement of the universe's density and compare it to the critical density, we could learn our fate. We will look at that next. Afterward, we will consider how other issues, like dark energy, change the entire picture.

18.2 The Density of Our Universe

Unfortunately, while there are many ways to measure the density of the universe, most of them are extremely difficult. Early attempts focused on trying to measure the number and types of atoms both in and in-between galaxies. These scientific efforts went something like this:

From the amount of light we see from the stars, we can get a good idea of the amount of mass of all the stars in a galaxy. By looking at how this light interacts with atoms between the stars and our eyes, we can also get a measure of the mass of all the stuff in-between the stars. Utilizing the most sophisticated methods of measuring the visible mass in the universe, scientists' best estimate of the amount of mass in the photons, atoms, and neutrinos total a few percent or so of the mass needed to give the universe a critical density.

Different methods are needed to determine the amount of mass from dark matter. From galaxy rotation, lensing, and other methods we have discussed we see that galaxies have much more mass in dark matter than in atoms, and there is significant dark matter between galaxies. Typical estimates using these methods indicate there is about five times more dark matter in the universe than in the known particles. The combination of these two types of measurements, and others, tells us that the amount of mass in the universe totals up to less than 30 percent of the critical density.

A completely independent way to measure the density of the universe all at once uses the temperature variations in the cosmic background radiation (see Figure 18.3). If you look closely, you can see that there are regions where the temperature is either higher or lower than average. If you think of these regions as spots on a map, it is straightforward to measure the size of the spots (how wide they are) and use that measurement to indirectly determine the density of the universe. A more correct way to say it is that we are measuring the overall curvature of the universe which is related, through general relativity, to the density of the universe.

To understand how we measure the density from the size of the spots, we note that the true distance across the spots is known for reasons which are beyond the scope of this book. What we are doing is measuring the size of the spots after the light from them has traveled through the universe since recombination (like we did with gravitational lensing in Chapter 6). Since the overall curvature of space-time in the universe is determined by the density of the universe, the direction of the light is curved as it travels through the universe as shown in Figure 18.3. In a universe with a density that is greater than the critical density (indicating large

NASA / WMAP Science Team

FIGURE 18.3 The top plot shows temperature fluctuations in the cosmic background radiation. These tiny fluctuations in the measured temperature distribution appear in different directions in the universe. The middle row shows how light from high (or low) temperature spots travel through the curved space-time of the universe, and reach us today. On the bottom-left is a simulation of what would happen to the light from a spot as it travels through a universe that will end in a big crunch. On the bottom-right is one that will expand forever. Between them is what would happen to light in a universe with a critical density (overall zero curvature of the universe). Since we know what the true width of a spot should be, we can use computer simulations to do a full prediction of the cosmic background radiation in the three different density scenarios. The middle simulation most closely resembles the data. To get a sense of the size of the spots in the middle plot, consider the size of the Moon as seen from the Earth. That size would cover about the diameter of a spot. (Color version on page C-11.)

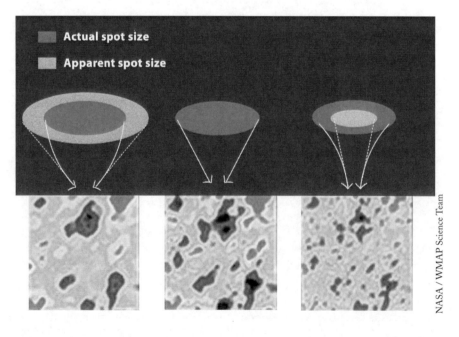

NASA / WMAP Science Team

BIG BANG, BLACK HOLES, NO MATH

amounts of mass in the universe), we predict that the size of the spots will appear to be larger than expected. On the other hand, if the density is less than the critical density, then we expect the size of the spots to be smaller than expected. If the density of the universe is equal to the critical density, the measured distance would show as they are. Using computer simulations we can show what the temperature distribution of the cosmic background radiation look like on the full sky. These are shown in the bottom of the figure.

Comparing these predictions to observed data, our findings point to a universe with 100 percent of the critical density. A naïve interpretation of these two density measurements—comparing the mass we can measure to the result from the cosmic background radiation—would appear inconsistent.

We next describe a third way of measuring the density of the universe. By looking at the distances to the most distant galaxies, we can learn about the expansion of the universe over time.

Even though the universe is expanding, we expect the speed of the expansion to change over time because of the mass in the universe. Think back to our ball thrown into the air. As it moves away from us, we can measure both where it is and how fast it is going. A short time later, it has a new location and its speed has changed. Since very little time has passed, the ball may still be moving up into the air, but it would have slowed down because of the acceleration (or pull, or force) of gravity. This slowing down is called a **deceleration**.

General relativity predicts that in the same way that a ball slows down because of the pull from the Earth's mass, the expansion of the universe should also be slowing down—or decelerating—because of the mass that fills the universe. Whether or not this expansion ever comes to a complete stop, however, ultimately depends on the universe's density. The larger the density of the universe, the more slowing down we would see when we look at the speeds of the most distant galaxies, which should have been decelerating over billions of years.

As detectives, we have two density measurements: one says the universe has less than 30 percent of the critical density, and one says it has 100 percent. The expectations for a universe with critical density predict that the most distant galaxies should have slowed down more over time than if the universe has less than the critical density. Perhaps this method can tell us which of these two hypotheses, if either, best describe the data.

To be able to see—and study—the galaxies that emitted light more than ten billion years ago (so we can see how the expansion of the universe has changed over time), we need to find ones that are emitting enormous amounts of light. As described in Chapter 16, supernova explosions are incredibly bright, and thus you can see them from the most distant reaches of space. Fortunately for scientists, we actually know how many photons are being emitted by one special kind of supernova known as Type Ia when it is at its brightest.

As described in Chapter 16, Type Ia supernovae occur when a white dwarf essentially "eats" part of another star. This can occur, for example, when it collides with another star, or if it is part of a binary star pair. The white dwarf can only eat so much until it reaches a certain mass (like for high-mass stars as discussed in Chapter 16), and then explodes its contents into space. Since it explodes when it reaches this critical mass, the amount of energy churned out in any explosion of this type will always be similar and from the number of photons we observe, we can tell how far away it is. Since a supernova explosion is very violent, we can see it in the furthest reaches of the universe, and we can measure how long ago it exploded so that we are just seeing it now.

In a sense, we can use these types of supernovae like we did the Cepheid variables we described in Chapter 10: if you find a Type Ia supernova in a distant galaxy, you can figure out the distance to that galaxy. By measuring the red-shift of the spectral lines, we can also measure the stars' original speed away from us. With data from enough supernovae, we can measure how much the universe is decelerating and determine the fate of our universe based on the predictions of general relativity. In a very real way, this data should allow us to measure the density of the universe.

We are now ready for a comparison.

18.3 AN ACCELERATING UNIVERSE?

When scientists in the 1990s used supernovae to study the distant galaxies, much to their surprise, they found not only that the universe was not decelerating, but also that it was actually accelerating. It is as if we threw a ball into the air and it slowed down for awhile, but when it got high enough, it actually began to speed up. We would not be surprised if a bird did this, but a ball? Nevertheless, it seems that at some point after the beginning of the universe, something started to speed up the expansion. But, what?

If the objects in the universe are accelerating away from each other, something must be causing this acceleration. Everything else in the universe that causes an acceleration has mass or energy. Since we do not "see" anything causing the acceleration, we say that the cause is "dark," like dark matter. But since matter only provides a gravitational attraction, dark *matter* can't be involved. Since this "something" has energy, but is not matter, we call it **dark energy**.

18.4 WHAT'S THE MATTER (AND ENERGY) IN THE UNIVERSE?

Using the speeds and distances of supernovae (as well as a lot of other information), we measure the amount of dark energy in the universe and find that it is the equivalent of some fifteen times the amount of regular matter. This means it accounts for about 70 percent of the universe's critical density.

For reasons that are beyond the scope of this book, the spots in Figure 18.3 also help us measure not just the overall density of the universe, but the fraction of the universe that is made of matter, and which fraction of that is in known particles, and what is in dark matter. The result, shown Figure 18.4, is that our best estimate is that about 5 percent of the critical density is from known particles (with atoms making up most of this 5 percent), about 23 percent of the critical density is in dark matter, and 72 percent in dark energy.

There are two remarkable results worth noting. The first is that we only know about 5 percent of what makes up our universe. Then again, when we look for the atoms (and other known particles), we are only able to observe between 0.5 percent and 2 percent, which means there are a lot of unaccounted-for particles. To be fair, it is very hard to measure the amount of gas in the universe, so we may have just not yet found them. The second remarkable result is that, assuming we find this missing mass, the entire story hangs together in that we observe a universe with the critical density and that we have an preliminary accounting of all the things in the universe that contribute to this density.

While it is still too early to say that the case is closed, an enormous amount of data is all pointing in the same direction. Then again, we do not know what dark energy *is*—all we have done is give it a name and describe it a little bit.

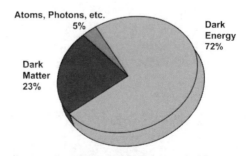

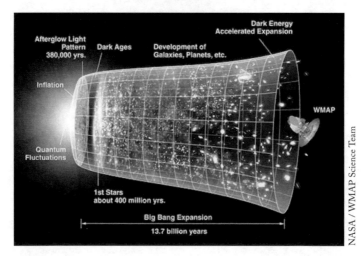

NASA/WMAP Science Team

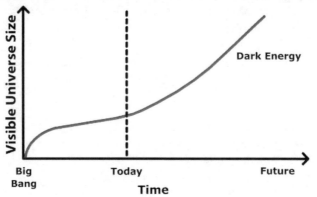

FIGURE 18.4 An accounting of the mass and energy in the universe, as well as two different views of its evolution and fate. Our best understanding today is that the universe will expand forever. The bottom figure is a graph that is similar to that in 18.2, but takes into account our best understanding of the universe, including dark energy. (Color version on page C-10.)

BIG BANG, BLACK HOLES, NO MATH

18.5 The Evolution of the Universe and its Fate

We are finally ready to describe our best understanding of the history and fate of the universe (Figure 18.4) after its earliest moments. As far as we can tell, the universe has always been expanding. For billions of years it was decelerating (its expansion rate was slowing over time) and during this time, protons, atoms, galaxies, stars, and planets formed.

About six billion years ago, the universe started to accelerate. In other words, the expansion started speeding up, as if someone hit the accelerator in a car. While we do not know how or why this occurred, we see that the combined effects came to give us the size and expansion rate we currently see. Today, dark energy causes the universe to keep expanding faster and faster.

Inquiring minds want to know what dark energy is, why it might be causing the universe to accelerate, and if we have a fundamental misunderstanding of the laws of physics. Scientists do not yet know the answers to these questions, but there are some interesting ideas that are worth mentioning that actually originated back in the early part of the twentieth century.

Before the first evidence that the universe was expanding, Einstein was trying to explain the universe using his theory of general relativity. This presented him with a challenge: his theory of general relativity had to account for the "observed fact" that the observed stars (remember the Milky Way was thought to be the entire universe back then) showed no evidence of a universe that was either expanding or contracting.

If there is a ball in the air, it has to be moving up or down—it cannot hover for more than an instant without something holding it in place. Likewise with the universe, something had to counteract gravity, so Einstein introduced a "trick" into his equations. He postulated that there was some unknown property of space-time he called a **cosmological constant**, which opposes the force of gravity and precisely balances the universe to be such that it is neither expanding nor contracting. While he did not have an explanation for where this property came from, it was consistent with the laws of general relativity, so he inserted it anyway to explain the data he had.

After it was shown that the universe was expanding, and Einstein's own general relativity could be used to explain this expansion, there was no need for a cosmological constant. Einstein allegedly called his idea for a cosmological constant his "greatest blunder."

What makes this story remarkable is that today's data are well-explained by this unknown acceleration of the expansion of space-time in Einstein's predictions. However, instead of it keeping the universe from expanding at all, it describes the acceleration of the universe.

Now, nearly a hundred years later, we can ask: Was Einstein right about there being a cosmological constant, albeit for the wrong reasons? Is there really some property of space-time that accelerates the expansion of the universe to be faster and faster?

We do not yet know the answers to these questions because we do not yet understand the nature of dark energy. Similarly, we do not really know the fate of our universe.

Dark energy could cause the expansion rate to continually accelerate, making space-time stretch so much that, eventually, our neighboring galaxies are so far away that we cannot see them anymore. If this is true, it means we live in a special time in history where we can still see galaxies and have the ability to study them to tell us about our universe. There are some versions of dark energy that are so extreme that they predict what is known as a **Big Rip**. In this case, the expansion rate would eventually become so rapid that it overcomes the gravitational attraction that keeps the galaxy together, then rips apart our Solar System, and then atoms, nuclei, and protons. This would be hundreds of billions of years in the future.

It's important to remember that Einstein's predictions for the acceleration of the universe just tell us *how* it accelerates. It provides nothing in terms of *why* this acceleration is occurring. Indeed, this acceleration confounds modern physics and has no explanation. It is not even clear that we are close to understanding it.

To make matters worse, it is always possible that another unknown force could emerge to make the universe shrink and end in a big crunch. Maybe we will end up somewhere in-between these scenarios—or experience some other future we have not yet envisioned.

To figure out the ultimate fate of the universe, we need to know the nature of dark energy and more. We clearly have entered into an exciting time in our understanding of the cosmos. Hopefully, it will not take too long to discover the true nature of dark energy and its role in our universe's future.

We next turn to the early moments after the bang and continue traveling backward to even earlier moments than that.

Particle Physics, Dark Matter, and the Very Early Universe

CHAPTER 19

It is not hard to see that any understanding of the universe and its evolution is intimately tied to our knowledge of particle physics.

Today, the universe is filled with atoms, but each is made of fundamental particles (quarks and electrons) that would have been created in the high-energy collisions of the insanely hot early universe. To learn more about times earlier than a millionth of a second after the bang, we need to know more about how these particles and others were created, interacted, and decayed over time.

Since there is some evidence that dark matter is a type of "not-yet-understood" particle, which would mean that it must have been created at some point in the history of the universe, we will include it in our discussion.

We will start this chapter with a more detailed description of the known particles (listed in Chapter 3) and take a more in-depth look at how they interact with each other. Our knowledge of the fundamental particles and their interactions is known as the **standard model of particle physics**, or just the "standard model," for short. While this description has been tested with remarkable amounts of data, it does leave many questions unanswered, most notably those dealing with dark matter. By looking at models *like* the standard model, scientists are designing extended versions of the theory that might provide an explanation.

One such version, called **supersymmetry**, has real advantages and, in addition, predicts a new fundamental particle that could be the dark matter particle. Although we have not yet been able to properly check this idea, we will talk about how it ties into the rest of our story and ways to test the hypothesis.

Indeed, the search for new particles is one of the most important detective cases that scientists are attempting to solve. We will conclude our chapter with a description of some of the giant experiments that are under way, and discuss why a major discovery might be just around the corner.

19.1 FUNDAMENTAL PARTICLES AND FORCES OF NATURE WE HAVE ALREADY DISCUSSED

The study of particle physics, in one form or another, is as old as the study of the heavens. The Greeks theorized that there were fundamental building blocks of nature, which they called atoms ("atomos" in Greek means "not cuttable" or "not divisible").

In the 1800s, scientists began to gather hard evidence that matter, like rocks and air, was made of different types of atoms. As time marched on however, we learned that these atoms are neither fundamental nor indivisible. Despite this, the name "atom" has stuck. As we have seen, each atom is an arrangement of nuclei surrounded by electrons that chemists have grouped into the periodic table.

The modern study of fundamental particles did not start until about 1900 with the discovery of the electron, which is still believed to be fundamental. In the 1920s, we learned that the nucleus was composed of protons and neutrons, which were then also thought to be fundamental particles. In the 1970s, however, scientists broke apart protons and neutrons and showed they were made of quarks. So far, no one has managed to break apart electrons or quarks, so there is no evidence that they are composite particles—although that possibility certainly exists.

In addition to looking at the many different types of fundamental particles, we have learned about the forces between them. Electromagnetism describes the attraction and repulsion between particles with electric charge. The particle that "carries" the electromagnetic force is the photon, and atoms are held together when the photon does the "talking" between the negatively charged electron and the positively charged proton. A Feynman diagram describes how these particles "talk" with each other, and describes their interactions (Figure 7.6). The interactions between quarks that hold protons together are similar, except that instead of using a photon, they use a gluon (Figure 8.1).

Particle physicists realized that matter particles, like quarks and electrons, behave similarly to each other in many ways, but differently than the force carriers. For this reason, matter particles are known as **fermions** (since they were first described by Nobel laureate Enrico Fermi) and force carriers are known as **bosons** (since they were first described by Satyendra Bose, who should have received a Nobel Prize for this work).

Gravity does not work well in this framework because there is no good description of gravity in particle physics or quantum interactions in general relativity. Scientists would say that we do not yet have a good quantum theory of general relativity. That said, scientists have theorized that a force particle of gravity exists (a boson dubbed a **graviton**). While physicists have hunted for direct evidence for the graviton particle, there is no convincing evidence one way or the other.

19.2 OTHER FUNDAMENTAL AND COMPOSITE PARTICLES: THE HEAVY AND SHORT-LIVED

The standard model of particle physics contains a list of particles (Table 3.1) and describes how these particles interact with each other. Since each type of particle would have been created in the high-energy collisions of the early universe, it is worth mentioning them by name.

There are six types of quarks, named **up, down, strange, charm, top,** and **bottom**. There are likewise six types of "electrons" (known generically as **leptons**): **electron, muon, tau,** and a **neutrino** corresponding to each of the three. All the quarks and leptons are fermions. There are also carriers for each known force: the gluon (carrier of the strong force), the photon (carrier of electromagnetism), and the W and Z bosons (carriers of the weak force), which we will talk about next. Each force carrier is a boson. The graviton, if it exists, would be the force carrier of gravity.

Each one of these particles has an anti-particle version. That is a lot of particles, which is one of the reasons people suspect they might all be composed of something "more" fundamental. Perhaps we are living in a time similar to the period when scientists filled in the periodic table, but before they figured out atoms were composite.

Though incredibly important, the weak force isn't well known, even though it is responsible for making neutrons decay when they are outside a nucleus. Some examples are shown in Figures 19.1 and 19.2. A Z-boson can be created when a high-energy particle and an anti-particle collide; in Figure 19.1, we see this happening with an electron and a positron. The Z-boson can then decay into a particle and an anti-particle, like a muon and an anti-muon, or a quark and an anti-quark. Such an interaction can start and finish in much less than 10^{-20} seconds.

We can now understand neutron decay (Figure 14.3) via the full Feynman diagram shown in Figure 19.2. The neutron is made of three quarks: two downs and one up. One of the down-quarks can turn into an up-quark and a W-boson, leaving us with two up-quarks, a down-quark, and a W-boson. But two up-quarks and one down-quark is just a proton (Figure 8.1). So a neutron has decayed into a

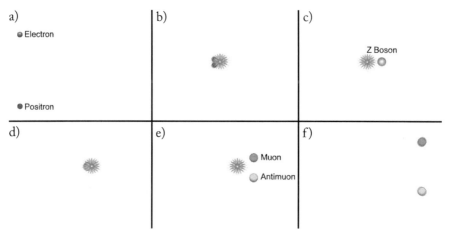

FIGURE 19.1 A set of action shots showing some weak-force interactions that are important in particle physics and in the early universe. The top set shows an electron and a positron that collide and turn into a Z-boson. The Z-boson then decays into a muon and an anti-muon. In the bottom set is the same scenario, but here, the Z-boson decays into a quark and an anti-quark. These are but a few of the possibilities.

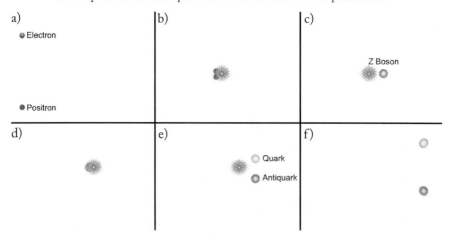

proton and a W-boson. The W-boson would then rapidly decay into an electron and a neutrino, which we can see in experiments. From the outside, we would see a neutron decaying into a proton, an electron and a neutrino; now we know what is going on inside.

In the bottom of Figure 19.2, we see how a high-energy electron and a proton interact to create a neutron, as in the formation of neutron stars (Chapter 16). The interactions with a W or Z boson are also central to the collisions described in Chapter 13 and helped produce the vast array of particle types in the early universe.

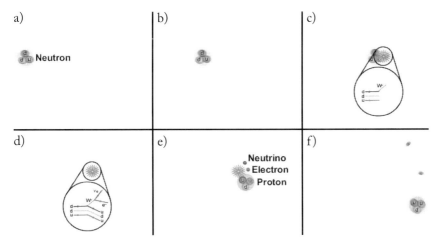

FIGURE 19.2 Weak interactions that were important in the early universe, as well as in the formation of neutron stars. The top set of action shots explains the decaying process of a neutron and shows more detail of what was occurring in Figure 14.3. The bottom set of shots illustrates the joining of an electron and a proton to create a neutron, as occurs in the formation of neutron stars. Note the Feynman diagram of the interactions in the insets. (Color version on C-12.)

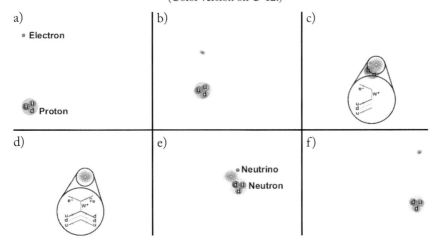

Composite particles are also an important part of the early universe. As we have seen, when three quarks are put together in the right way, we have neutrons and protons. We can also put electrons and protons together to create atoms. Can we put each of the fundamental particles together in different ways to create other composite particles? The answer is yes and no. Some combinations will work, but even for the ones that do, most do not live very long.

Scientists started discovering these short-lived composite particles in the 1930s, but did not understand their nature until the 1970s, with the discovery of quarks. There are many ways to put quarks and/or anti-quarks together to make composite particles. Those made of quarks and/or anti-quarks are called hadrons. A good example of a hadron is a proton. This helps explain the name of the world's biggest "atom smasher," the Large Hadron Collider—or LHC, for short.

Since the LHC takes the protons, and accelerates and then collides (smashes) bunches of them into each other, our accelerator is called a "collider." Finally, as it is about seventeen miles in circumference, it is definitely large. Hence the name Large Hadron Collider!

Anti-particles can also exist in composite particles. For example, two anti-up quarks and an anti-down quark can exist as an anti-proton as shown in Figure 19.3. A quark and an anti-quark can even exist as a composite particle for a short time. Different combinations of up/anti-up and down/anti-down quarks produce a particle known as a **pion** (from the Greek letter π, pronounced pie-on). Various combinations of strange or anti-strange quarks paired with up, anti-up, down, and anti-down quarks create a **kaon** (from the letter K, pronounced kay-on).

It is odd to call these composite objects "particles" because they decay almost immediately, often as little as 10^{-24} sec. Then again, before they decay, they act like all the other particles we know.

The bottom line is that these particles were vitally important in the very early universe, and we need to know more about them if we want to move closer toward understanding what was happening shortly after time began. We know these particles are not around anymore because if they can decay, they have already done so.

Ultimately, to understand the universe at times before a millionth of a second after the bang, we have to know which particles can exist in nature, and how they interact with each other. To understand the evolution of the universe, however, we also need to know how and when particles decay.

Sad to say, we do not yet know why there are so many different fundamental particles. We do not know whether they are truly fundamental, nor if we have a complete list. There may even be other forces and/or particles we have not yet discovered. These are the kinds of issues that fascinate scientists for many reasons, only one of which is that they help determine what happened in the very early universe. The hunt for new particles is alive and well.

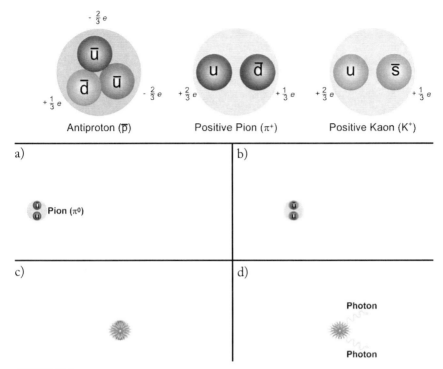

FIGURE 19.3 An artist's conception of the particles known as an antiproton, a pion, and a kaon (ignore the shadings and the sizes of the particles). An up-quark and an anti-down-quark can be combined into a pion with a positive charge. The combination of an up-quark and an anti-strange-quark is known as a kaon with a positive charge. Most pions live for only about 10^{-24} seconds before decaying. In the bottom two rows, we see a set of action shots showing a neutral pion decaying into two photons. This explains why there are not many around today. During the very early universe (especially when it was less than 10^{-24} seconds old), however, they would have been important.

19.3 COLLIDING CLUSTERS OF GALAXIES PROVIDE EVIDENCE THAT DARK MATTER IS ACTUALLY LOTS OF PARTICLES

Scientists have spent years trying to figure out what dark matter actually is. Could it be material, like planets that do not shine, much like our Earth? Brown dwarfs? Black dwarfs? Could dark matter be large amounts of a single type of fundamental particle? What clues do we have?

While any of the above ideas could explain lots of mass we don't see interacting with light; dark matter made up of large, dense objects like stars and planets would not bend light the way we observe in galaxies.

If dark matter is a particle, it cannot have a positive or negative charge because it would interact with light and we would see it. It also cannot interact much with regular matter, like atoms, because there are dark matter particles in our galaxy and we would have seen atoms bouncing off "nothing" here on Earth. These clues are not much to go on, but it suggests looking for a particle that has mass, has neutral charge, and doesn't interact much.

One way to test this hypothesis is to point telescopes in a direction where lots of dark matter should be colliding and "watch" it in action. Instead of considering collisions of galaxies, let us consider colliding clusters of galaxies, which are far bigger (roughly ten million light-years across, which is about a hundred times bigger than our Milky Way). That is a lot of dark matter.

For simplicity, let us start by considering each cluster as a simple combination of intermingled atoms and dark matter. What do intermingled atoms and dark matter do in collisions? To answer that question, we first consider atoms and dark matter independently (Figure 19.4).

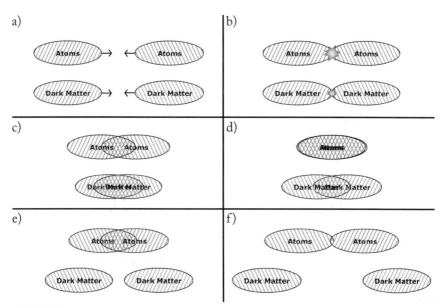

FIGURE 19.4 A series of action shots showing matter in motion and on a collision course. In the top part of each pane are two groups of atoms moving toward each other. They will slow down as they pass through each other and interact. In the bottom part of each pane are two groups of dark matter on a collision course. They will not slow down much as they pass through each other because they interact only through gravity. Thus, the dark matter keeps going at basically the speed it was before the collision. Note that the dark matter figures are our best guess as to what will happen with dark matter because we do not yet understand its true nature.

Atoms in these clusters should collide as they do here on Earth. Some atoms will collide like billiard balls and move backward, and some will only strike a glancing blow. On average, both groups of atoms will slow down as they pass through each other. I like to think of them as two water balloons thrown at each other, then meeting with an impressive "splat!" in mid-air.

Dark matter particles, on the other hand, would not interact a great deal, except gravitationally. This means the dark matter portion of the cluster would not be slowed down very much in a collision. It would be more like two ships passing in the night. With this in mind, we expect two colliding sets of combined atoms and dark matter to look like Figure 19.5. As time goes by, the dark matter portion of a cluster of galaxies will separate out after the collision.

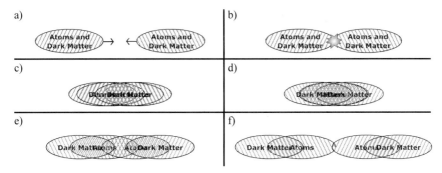

FIGURE 19.5 A series of action shots showing clusters of galaxies made up of atoms and dark matter on a collision course. Each has a distinctive way of interacting when it collides. The atoms will move through each other slowly because of their interactions and the dark matter will pass through quickly. During the course of the collisions, the two parts of the cluster will separate. As in Figure 19.4, the dark matter figures are our best guess as to what will happen with dark matter because we do not yet understand its true nature. However, there is data that supports this hypothesis in Figure 19.6.
(Color version on C-13.)

We can observe a snapshot in time of two real clusters of galaxies that have already collided using a combination of observational techniques: gravitational lensing for the dark matter (described in Chapter 6), and regular telescopes for the light from the interacting atoms. Using data taken in 2006, Figure 19.6 shows a composite image of two colliding clusters. The blue light represents the amount of matter present as measured from lensing (dark matter), and the red light represents the amount of atomic matter as measured from observed light. The clear separation between the dark matter and the atoms suggests that dark matter consists of particles that do not interact much and that they have traveled through the cluster and away from the rest of the atoms.

FIGURE 19.6 Real data showing a study of two colliding clusters of galaxies. The image's blue region shows where the dark matter is (using gravitational lensing) and the red region shows where the atoms are (using light from the collisions). This data is interpreted as two colliding clusters of galaxies, each of which started as a collection of atoms and dark matter. This set of observations could be explained by the pictures in Figure 19.5. (Color version on page C-14.)

19.4 SUPERSYMMETRY AND DARK MATTER: FUNDAMENTAL PARTICLES WE HAVE NOT YET DISCOVERED?

Since there is no evidence that we have discovered all of the fundamental particles—and there are ample reasons to believe there may be others waiting to be discovered—as good detectives and scientists, we continue the hunt for new particles. In our quest, we need other clues—and good clues are hard to come by.

Every so often, detectives try to take two separate cases to see if they have some similarities; maybe the two were perpetrated by the same criminal. If so, then by looking at both together, perhaps a new connection can be made and there is a better chance of solving the mystery. This does not always work—in fact, it rarely works—but sometimes it is our only hope. For this reason, scientists have been trying to put together the dark matter puzzle with the unexplained-differences-among-the-various-types-of-particles puzzle. We will start with more on the latter puzzle, and then show how the two puzzles might be related.

For many years, scientists have noticed that some particles are very similar to each other and some are very different. They also noticed that sometimes they have partners and sometimes they do not. For example, an electron (which is matter) has an anti-matter partner. These two particles are, in some ways, mirrors of each other. Some of their properties are identical and some are opposites.

Specifically, electrons and anti-electrons (positrons) have the same mass, but opposite charges. Fermions—particles of matter like electrons and quarks—have many characteristics in common with each other. Likewise, bosons (force-carrier particles like photons and gluons) have certain resemblances. But the two groups behave differently from each other. Why is this?

Some theories, like supersymmetry (or SUSY, for short), predict that in the same way that each particle has an anti-matter partner, each type of matter particle must have a force-carrier-like partner, and vice versa. These fundamental "mirror" particles may exist in nature, but we have not yet discovered them.

In their optimism, scientists have already given these "super"-partners names: an electron's superpartner would be called a supersymmetric electron. Each of the particles of the standard model listed in Table 3.1 has a SUSY version. Collectively, all these supersymmetric particles would be called **superparticles**. Each superparticle would also have an "anti-superparticle." Huge numbers of these types of particles would have been created in the high-energy collisions of the very early universe and thus played an important role in its history. Where are they today?

Before answering this, we need one more piece of information about SUSY. There is reason to believe that if superparticles can exist, then the amount of "super-ness" in an interaction, like the amount of charge, must stay the same over time. To understand what we mean by this, let us take another look at electric charge.

As we mentioned in Chapter 7, if there is an interaction between two particles—or if a single particle decays—the amount of electric charge cannot change from before the interaction to after the interaction. For example, if I have a neutron (which is neutral), it can decay into a proton (positive), an electron (negative), and a neutrino (neutral) because the combined charge at the end is zero—the same charge it originally carried. We say that the amount of charge has been conserved in the decay.

An example of this interaction—but with superparticles instead of standard model particles—is shown in Figure 19.7. It starts with one supersymmetric particle, a supersymmetric electron, which decays into an electron and a supersymmetric photon. To understand how this is allowed, we point out that all regular particles have a super-ness of +1, and SUSY particles have -1. What is different is that to find the total amount of super-ness, we do not add the super-ness of the particles involved, but multiply them. So, in our example, the amount of super-ness begins with a value of -1 (there was only one particle and it had a super-ness of -1). At the end, we had two particles, one with a +1 and one with a -1, so multiplying them we get (-1)*(1)=-1, so the super-ness at the end is -1. Thus, the amount of super-ness before and after the interaction is identical; super-ness has been conserved. In the same way that a pair of photons (both neutral) can collide and create an electron and a positron (net charge of zero), a collision of high-energy particles could create a superparticle and an anti-superparticle.

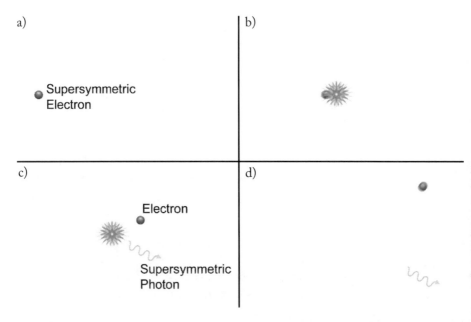

a) Supersymmetric Electron

b)

c) Electron
Supersymmetric Photon

d)

FIGURE 19.7 The decay of the hypothetical particle known as a supersymmetric electron into an electron and a supersymmetric photon (another hypothetical particle). Before and after the decay, there is one and only one supersymmetric particle. If this version of supersymmetry is correct, the lightest supersymmetric particle would never decay and could be dark matter.

If the super-ness is conserved, it has enormous implications. Specifically, the lightest supersymmetric particle could *never* decay.

A neutron decays because the proton is lighter and because it *can* decay into a proton. A proton, on the other hand, cannot decay because there is nothing lighter for it to decay into that is allowed by the known forces of nature. The same is true for electrons.

The bottom line is that if the lightest supersymmetric particle also cannot decay, it could last forever. So what? If the lightest supersymmetric particle is also neutral (not positively or negatively charged), then it does not interact with photons. This, then, could prove to be the dark matter particle, and lots of them could make up the dark matter in the universe.

High-energy collisions in the early universe could have created vast numbers of supersymmetric particles, and all of them would have decayed except the lightest one. Since the lightest ones cannot decay, they are the only ones that would still be around today. They would still be out there, helping galaxies rotate and causing gravitational lensing.

Before we get too excited, though, remember that there is no direct evidence that supersymmetry is correct: we have not observed any supersymmetric particles in an experiment. As good detectives, we need to be thoughtful and reserve judgment. This lack of evidence does not mean they do not exist: they may simply be so heavy that our best "atom smashers" do not have enough energy to create or discover them. Einstein's $E=mc^2$ means that you need lots of energy to make any heavy particle.

Figure 19.8 shows pictures of the world's two biggest particle accelerators; today's version of the atom-smasher. These enormous devices accelerate particles to high energies and then bash them together in the middle of a giant particle detector. Through these high-energy collisions, scientists recreate the energies of the collisions of the early universe.[1] They may be our best bet to understanding dark matter, particle physics, and cosmology all in one experiment.

We quickly mention each. One is the Fermilab Tevatron, which is located just outside Chicago. This accelerator is almost four miles around, collects large numbers of protons into bunches, and accelerates them to high energies. It does the same for bunches of anti-protons. It then directs these bunches toward each other so that they collide in the center of a giant particle detector (Figure 19.8). This accelerator, and the detectors used to study the collisions, discovered the top quark in 1996. As of today, it has stopped taking data, but scientists are still analyzing the data already taken.

The world's highest-energy accelerator is the LHC, located at the CERN laboratory in Switzerland. The LHC is very similar to the Tevatron, but it is more than

---

[1] Some people worry that they are recreating the big bang. Clearly, this isn't the case. At best, they are creating the energy conditions of the universe about a picosecond (10^{-12} sec) *after* the bang.

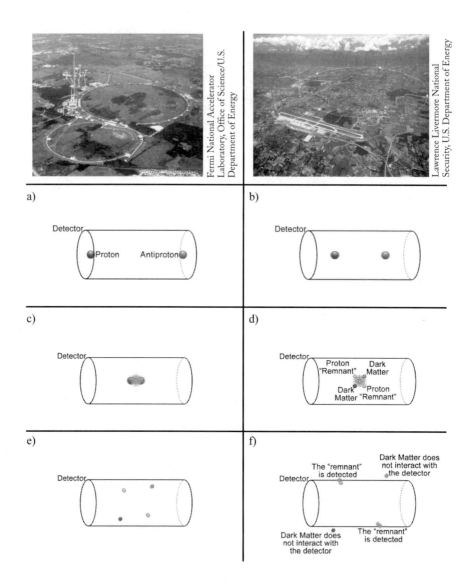

FIGURE 19.8 Aerial views of the Tevatron particle accelerator at Fermilab (top left, four miles around the bigger of the two circles in the picture) and the Large Hadron Collider (LHC) at CERN (top right, about seventeen miles around the circle). Each has created many high-energy collisions that might have produced dark matter particles. Scientists have studied these collisions to both look for dark matter and test supersymmetry. The bottom set of drawings show a simplified example of a high-energy proton and anti-proton colliding in a detector at Fermilab and creating dark matter particles that can be detected. (Color versions on page C-15.)

four times bigger in circumference and collides bunches of protons into other bunches of protons. As we write this book, its data-taking is well under way and scientists are excited because it can produce much higher energies than Fermilab. Also as of this writing, there is convincing evidence that the LHC has just discovered a new particle known as the **Higgs boson**.[2]

Figure 19.8 shows a simplified version of how dark matter particles could be produced in these high-energy collisions and then detected. A high-energy proton collides with a high-energy anti-proton and bursts them both apart. Part of the energy of the collision could create some dark matter particles, and the rest of the particles—the remnant—would hit the detector. Since the dark matter particles do not interact much (see Figure 19.6), they will just leave the detector.

In essence, we would simply need two things to discover the particle of dark matter: high-enough energy collisions to create dark matter particles and high-quality detectors to search for evidence of interactions where heavy, neutral particles left the detector without interacting. This is much easier said than done. Current experiments at the LHC have cost over $10 billion and thousands of scientists are needed. It is worth every penny.

Armed with our best understanding of what is happening in times earlier than a millionth of a second, we next move to times even earlier than that. We will discuss the idea of "inflation" in the initial moments after the beginning of the universe.

---

[2] The Higgs boson is a very important part of the standard model of particle physics, but its description is outside the scope of this book. See the Suggested Reading for more information.

Inflation and the Earliest Moments in Time

CHAPTER 20

Having completed our description of the fate of the universe, as well as a discussion of what the universe looked like a tiny fraction of a second after the bang, we conclude this book with a short and sweet chapter on our best understanding of its earliest moments.

This time in our universe's history is not as well-understood as the periods we have already discussed, but it is equally as exciting. Our goal here is not to present a detailed analysis, but rather to give you a taste of the amazing mysteries scientists are slowly unraveling.

So far, we have presented the big bang theory as a description of a universe that started with a minuscule size, followed by an expansion. However, there is good reason to believe that very early on, when our cosmos was about 10^{-35} seconds old, the expansion was not as straightforward as it is today.

In this chapter, we will present some of the evidence that suggests that the universe was very small for awhile (like during the first 10^{-36} seconds), then expanded at an incredible rate for a short time (doubled in size more than a hundred times over during the next 10^{-32} seconds), then settled down and evolved into what we see today. While a full story is far more complicated, scientists have dubbed this rapid expansion **inflation**.

We start by restating one of the most important pieces of data about our observable universe: the data is consistent with a universe that has a cosmic background radiation temperature that is virtually the same in all directions, but also has small fluctuations.

The reason this data (and much more) point to a period of rapid growth in the early universe is not obvious, so we will go the other way: we will describe why the data are *not* consistent with the simplest big bang model. Then we will describe a new version of the story, with inflation included, and demonstrate how it better explains our data. While we have good confidence that inflation did occur, we do not yet have a solid explanation of *why* it occurred.

20.1 THE UNIVERSE HAS THE SAME TEMPERATURE EVERYWHERE

The universe looks essentially the same in every direction (Chapters 10 and 12). Equally important is that the photons in the universe have energies that correspond to a temperature of 2.728 Kelvin. However, as shown in Chapter 15, there are variations of about 200 microKelvin (Figure 15.4).

Thus far, our explanation has been a simple big bang model. Specifically, the universe started as a single point in space-time and simply expanded and evolved according to general relativity. The uniform temperature was described as being the result of (at some point in the universe's history) the particles having banged into each other so many times that everything came into thermal equilibrium. Reaching thermal equilibrium at some point is what would make all the photons look the same in all directions and share a specific temperature. The variations were described as being due to quantum fluctuations in the early universe.

When we consider a simple big bang model, however, we find a significant problem. We can use relativity to figure out how fast the universe has been expanding at any giving point in time such that it started as a point and now has both the size and speed we see today. When we do this, we realize that the universe had to be expanding so quickly in the early universe that it did not have the opportunity to come into thermal equilibrium. This rapid expansion would not have allowed enough time for particles on one side of the universe to travel to the other side and bang into particles there. This would have been true even back when space-time was much smaller, regardless of the fact that the particles traveled at the speed of light.

As an analogy, think of putting both tap water and ice in a cooler. The water molecules near the ice will slam into the ice and get colder, and then collide with other tap-water molecules, making them colder as well. Likewise, the water molecules that hit the ice make the ice warmer, which eventually will melt the ice. It is just a matter of time before each molecule has collided so many times that everything in the cooler has the same temperature. Following this analogy, the explanation is that our universe came into thermal equilibrium because it had time to come into thermal equilibrium.

Next, consider what would happen if the cooler is expanding during this process. Eventually the cooler gets so big that all the water (and ice, if any) in the cooler ends up as puddles on the bottom of the cooler. Each puddle will have its own temperature since the temperature of one puddle no longer affects the temperature of any other since they aren't touching.

If we have just put the ice and tap water into the cooler and the cooler is expanding, we have two possible outcomes. If the cooler expands slowly enough that all the ice melts and everything comes into thermal equilibrium, then all the water comes to possess the same temperature. If this happens, then when the water divides into puddles, the puddles will all have the same temperature. However, if the expansion happens so quickly (say before all the ice melts), then some puddles will have ice in them and some will not; all the puddles will have a different temperature. So, either the expansion is slow enough that the water comes into thermal equilibrium and all the puddles have the same temperature, or it is so fast that the temperatures are different everywhere.

In a quickly expanding universe, the particles are like the water molecules; if they never come into equilibrium, then the universe will not have the same temperature throughout. The calculations of a simple big bang model predict that the universe expanded so quickly that it would not have had time to come to thermal equilibrium.

So did the universe come into thermal equilibrium a short time after it began? If not, how did the cosmic background radiation come to look almost the same in all directions? How did every part of the entire universe get to be so close to the same temperature if big chunks of the universe could not interact with each other? Why are there small variations in the temperature in different directions?

It is certainly possible that the universe exploded into existence in almost exactly the same way in all directions, but that does not explain why the energies of the photons look like they are from collisions that eventually produced thermal equilibrium.[1] Could it have been a really special explosion that makes the photons "look" like they came from almost perfect thermal equilibrium but with small variations?

While this is possible, it is important to consider the option that something else in the universe's history caused it to appear as almost perfect thermal equilibrium, with small variations. Our best guess is that exactly such a thing occurred in the early universe; we call this expansion "inflation."

---

[1] As described in Chapter 9, if particles come into thermal equilibrium, then, on average, they all have the same energy. However, they are not all exactly the same energy—some will be slightly higher and some will be slightly lower. If they are all exactly the same, then it is not thermal equilibrium. The energies clearly look like they have a temperature.

20.2 INFLATION

Inflation describes the idea that in the very early universe, there was a sudden expansion of space-time far more extreme than the one predicted by the simple big bang theory. And, yes, the name "inflation" originates from the economics term describing the value of money. Let's start with what we mean by inflation.

When you put money in a bank account, you often accrue interest, say one percent per year. Thus, the amount of money you have will slowly and regularly expand. However, if the country you are living in has horrible inflation, the interest rate might be one hundred percent per day. In that case, the number of dollars in your account will have doubled after the first day, quadrupled after the second day, and quickly grown at an enormous rate thereafter. It is this rapid type of expansion we are envisioning when we describe the inflation theory of our universe.

No one knows for sure when inflation started, how long it lasted, why it stopped, or even how much the universe expanded during this time. However, we have many versions of the story that are consistent with the data and can explain a great deal.

To simplify the description, think of the universe as not expanding very quickly right after the bang (say 10^{-36} seconds or so after the beginning). During this time space-time would have remained very small. It then expanded at an incredible pace for a short amount of time (say 10^{-32} seconds). The expansion would have been very large and, while no one knows for sure how much, some estimates say the smallest amount the universe could have grown is by 10^{26}, while others estimate it could be as large as 10^{80} times its size.[2] This is basically doubling in size somewhere between one hundred and three hundred times. While these are vastly different numbers, the details do not really matter for our purposes. What matters is that after inflation was over, the universe continued to grow, but at a rate consistent with what we observe today. An artist's conception of this expansion is shown in Figure 20.1.

To understand why inflation explains the data, we consider what happens before inflation and what happens after inflation.

In the time before inflation, what is now the visible universe was very small. For reasons we are not exactly sure of, it came to become the same everywhere. It may well be that the universe was like the water in our cooler coming into thermal equilibrium; the universe remained small for a long enough time that everything

---

[2] Remember, nothing can travel faster than the speed of light. However, general relativity does not place a limit on the speed at which space-time can expand. Growing from the size of a single atom ($\sim 10^{-10}$ m) to the size of the Earth ($\sim 10^{7}$ m) is a factor of 10^{17}.

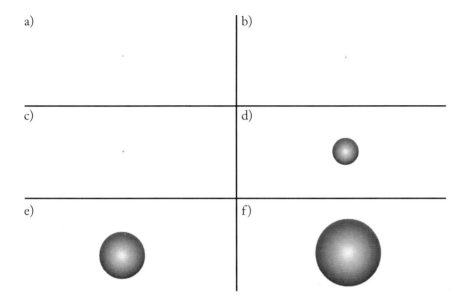

a) b)

c) d)

e) f)

FIGURE 20.1 An artist's conception of inflation in a few action shots. The universe remains small for awhile (a-c), then expands quickly (from c to d), then expands slowly again. These pictures are not drawn to scale.

in the universe came in contact with all the other things in the universe so that it came to be the same everywhere. While the analogy is helpful, we have to admit that we don't really understand the physics of these short times and small spaces so we don't know if there were particles before inflation (or if they interacted to come into thermal equilibrium). What is important is that, before inflation, the small amount of space that was the universe came to be effectively the same everywhere.

Next we consider what happens after inflation. After inflation the universe is so big that all the parts are so far apart that they don't all interact any more. Only things that are close to each other can interact regularly. However, since all the parts of the universe that are now far apart started the same way, the universe looks the same in all directions. Coming back to our analogy, if the water in the cooler is the same throughout and then expands rapidly, all the puddles at the bottom will be the same. They may not have started as having a temperature distribution, but after they become puddles, the water molecules in the puddle can interact so much that each puddle comes to be in thermal equilibrium with a specific temperature. Since all the puddles start the same, they all end up with the same temperature. So, in the same way that all the puddles of water at the bottom of the cooler have the same temperature, all the photons in the universe today have same temperature in all directions.

It's not that the universe is in thermal equilibrium now. It isn't. The universe we have looks the same in all directions because it started as a universe that was the same in all directions, and then came to have a temperature distribution due to lots of interactions between the particles in the universe.

20.3 Inflation and the Seeds of Galaxy Formation

While inflation explains why the universe shows a photon temperature that is basically the same in all directions, it can also explain why we see the quantum fluctuations in the cosmic background radiation. As described in Chapter 15 (see Figure 15.4), these fluctuations produced the seeds of galaxy formation.

To understand why this is, we go back to quantum mechanics. Before inflation, the universe came to be roughly the same everywhere. The reason it wasn't exactly the same in all directions is that because when things are very small, like the size of an atom, the effects of quantum mechanics are very important. For example, in an atom it affects how close electrons can be to protons when they orbit. Thus, the universe before inflation was almost exactly the same everywhere but had quantum fluctuations.

During inflation, these small differences would then be expanded along with space-time, and this is how they got to be spread around the universe. Said simply, the quantum fluctuations when the universe was small are there now on a large scale since the universe expanded so quickly. Going back to our cooler analogy, the water in the cooler was almost the same everywhere, but had small variations, but now that we have puddles, the temperature of the puddles are almost exactly the same but have small variations because of the variations when the water was all together and interacting all the time.

We now see these quantum fluctuations very clearly on a large scale as temperature variations in the cosmic background radiation. Indeed, they provided the seeds for the galaxies in the universe. Said differently, inflation explains why we have galaxies, and without inflation life as we know it wouldn't exist today.

While ideally we would end this discussion by telling you what caused inflation, unfortunately, we do not really know yet. Like with our description of dark energy, there are many ideas, but they are beyond the scope of this book. They are, however, discussed in some of the books in the Suggested Reading.

20.4 FINAL THOUGHTS

Our journey together has come to an end. I hope you agree with me that we live in an exciting time to be a human being. It is an era of great anticipation and discovery. If history is any guide, there are even more surprises in store for us down the road. A thousand years ago, many people thought the Sun moved around the Earth. A hundred years ago, most considered our Milky Way the only galaxy in the universe. Just twenty years ago, we had no inkling that dark energy even existed.

Our view of both the history and ultimate fate of our universe has radically changed in our lifetimes alone. Even as this book is being written, amazing new data pours in. Of course, as good detectives and scientists, we should be wary of any new clues. Current interpretations should be taken with a grain of salt.

In a few years, check with your bookstore for an updated version of this book to see how it all turns out!

Index

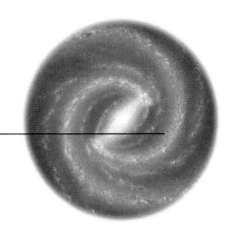

C

D

E

Earth
   age of, 127
   as center of universe, 117–118
   distance from Sun, 44
   formation of, 160, 189
   orbit of, 11–12, 26, 56–57, 59, 74
   rotation of, 27
Einstein, Albert, 111
   cosmological constant and, 213–214
   general relativity theory of, 29, 32, 52–57, 213
   quantum mechanics and, 71
   special relativity and, 52–53
   on speed of light, 43
Einstein ring, 61, 63, C–5
electric charges, 68–70, 86, 87, 216
electric field, 77
electromagnetic force, 65, 75–78
electromagnetism, 65, 68–71, 75–79, 85–87, 159, 216
electrons, 17–20, 22, 41, 68, 86
   anti-electrons and, 104, 225
   discovery of, 216
   in early universe, 149–150, 155
   electromagnetism and, 68–71, 85
   energy states of, 74, 79–81, 88–90
   interaction between photons and, 75–83
   quantum mechanics and, 71–75
   types of, 217
   wave-like properties of, 71–74
elliptical galaxies, 163, 165, 170
energy
   conservation of, 75, 79
   dark, 159, 211–214
   kinetic, 42, 43, 73
   from nuclear reactions, 90–93
   of photons, 80–83
   potential, 73
energy levels/states, 74, 79–81, 88–90, C–6
epicycle, 26
equatorial plane, 170

escape velocity, 193–194
event horizon, 196
evidence
   for big bang, 109–110, 125–138
   for black holes, 195–199
   scientific, 3, 27
evolution of the universe, 125, 139, 213–214, C–10
experiments, 32
explosions, 114–116

F

falsifiability, 30
Fermi, Enrico, 216
fermions, 216, 225
Feynman, Richard, 4, 77
Feynman diagrams, 76, 77, 78, 79, 86, 216, C–2
force, 50
force carriers, 216–217, 225
force particles, 21
fundamental building blocks of nature, 20–21, 66–68, 216
fundamental particles, 19–22, 41, 127–128, 144, 216–217, 220, 224–229
   See also specific types
fusion, 92

G

galaxies, 163–173
   appearance of, 164–165
   beginnings of, 170–173
   black holes in, 194–195, 199
   clusters of, 14
   collision of clusters of, 221–224, C–13, C–14
   collisions between, 170
   dark matter in, 59–64, 164, 168, 207

kinetic energy, 42, 43, 73
Kirshner, Robert P., 9

L

Large Hadron Collider (LHC), 144, 149, 220, 227–229
Lemaître, Georges, 114–115, 118
leptons, 21, 217
light
    See also photons
    atoms and, 79–83, 85
    colors of, 88–90
    dark matter and, 61–63
    Doppler effect and, 45–47
    electrons and, 75–79
    in general relativity theory, 57–59
    gravitational lensing of, 58–59
    infrared, 41
    as particle, 41–42
    red-shifts of, 122–123, 137
    speed of, 43–44, 52–53, 74–75, 194
    from stars, 90–94
    ultraviolet, 41
    visible, 40, 41, C–3
    as wave, 37–41
light-years, 43–44
liquid, 66
lithium, 154
lunar eclipse, 28

M

Mars, 12, 24, 25
mass
    of black holes, 198, 199
    conversion to energy, 57, 92, 131
    critical, 204
    in galaxies, 93
    gathering of, 159–160
    solar, 182

of stars, 182–183, 187–188
mathematics, 49n1
matter, 21
    dark. *see* dark matter
    structure of, 66–68
Maxwell, James Clerk, 43
Mercury, 12, 24, 59
meters (m), 9
metric system, 9
microwaves, 41, 109, 135–137, C–7
Milky Way, 13, 14, 111, 112, 163, 164, 168, 170, 199, C–1
models, 29, 30
molecules, 66
Moon
    formation of, 178
    orbit of, 11, 26, 50–51
    phases of, 28
muons, 22, 144, 146, 147, 148, 151, 217

N

nature, fundamental building blocks of, 20–21
negative charge, 69, 87
Neptune, 12, 59
neutral charge, 69
neutral particles, 69, 159
neutrinos, 21, 91, 92n4, 217
neutron decay, 155, 188, 217–219, 226–227
neutrons, 20, 68, 85, 88, 91, 155, 216, 218–219
neutron stars, 176, 183, 188, 189, 194, 218–219, C–12
Newton, Isaac, 24, 29, 32, 49–51, 56
nitrogen, 88
nuclear interaction. *See* nuclear reactions
nuclear physics, 20, 85–94
nuclear reactions, 90–93, 178–181, 182
nucleus, 17–20, 68, 86, 88, 155, 216

O

Occam's razor, 27
orbits
    of electrons, 71–75
    planetary, 10–12, 29, 56–57, 59, 74
oxygen, 88, 90, 93, 155, 160

P

particle accelerators, 144, 220, 227–229
particle physics, 20, 215–229
particles, 98
    about, 19–22
    changing between types of, 104
    charged, 71, 75–79, 87, 159, 216
    combinations of, 128–130
    composite, 41, 127–128, 144, 219–220
    dark matter, 215, 221–229
    decay of, 22, 128, 134, 146–150, 217–220, 225–227
    fundamental, 144, 216–217, 220, 224–229
    gravitational attraction between, 159, 160
    interactions between, 127–134, 144–149, 159–160, 215–221, 225–229
    light as, 41–42
    search for new, 215, 224–229
    in thermal equilibrium, 101–105
    in universe, 125–138, 144–151, 155, 215–229
Pauli exclusion principle, 185n3
Penrose, Roger, 199
Penzias, Arno, 136
periodic table of elements, 88
photons, 41–42, 65, 217
    See also light
    collisions between, 103–104, 144

    creation of, 104–105, 132
    in early universe, 149–151
    energy of, 80–83, 127, 130–131
    interactions with atoms, 75–83, 130–131, 156–159
    interactions with electrons, 75–83
    microwave, 136–137
    in space-time, C–8
    temperature and, 96–97
    in thermal equilibrium, 125
    in universe, 109, 135–137, 156–157
    virtual, 77, 79
    wavelengths of, 135, C–7
physics
    atomic, 17–18
    nuclear, 20, 85–94
    particle, 20, 215–229
pions, 220, 221
planets, 24
    formation of, 160, 178
    orbits of, 10–12, 29, 56–57, 59, 74
    retrograde motion of, 26, 27
    speeds of, 59–60
Pluto, 12
polio virus, 17, 18
positive charge, 69, 87
positrons, 21, 22, 91, 104, 149–150, 155, 225
potential energy, 73
protons, 18–20, 22, 41, 68, 70, 85–88, 147, 216, 220
Proxima Centauri, 13, 44
Ptolemaeus, Claudius (Ptolemy), 24, 26, 32

Q

quantum fluctuations, 171–173, 236
quantum mechanics, 5, 18, 35, 42, 49, 65, 71–83
    black holes and, 199–200
    in early universe, 171–173

W

water, 66
wavelengths, 38, 40, 42, 71–74, 94, 135, C–3, C–7
waves
  light, 37–41
  properties of, 39
  sound, 44–46
W bosons, 217, 218
weak force, 87n2, 217–219, C–12
white dwarfs, 176, 183, 184, 186–187, 210
Wilson, Robert, 136

X

X-rays, 41, 199

Z

Z bosons, 217, 218
zodiac, 24–26

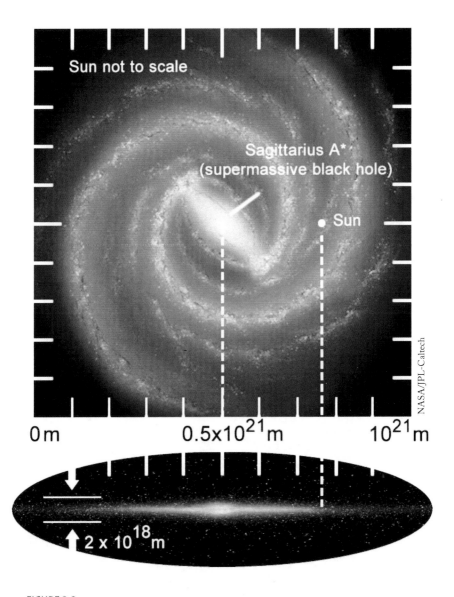

FIGURE 2.3e The top-down view and a side view of our galaxy. Note that it has many features in common with our Solar System, and that we are only showing the stars and not the dark matter.

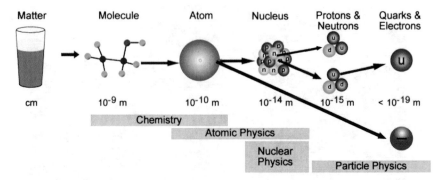

FIGURE 3.2 BOTTOM: From the size of atoms all the way down to quarks.

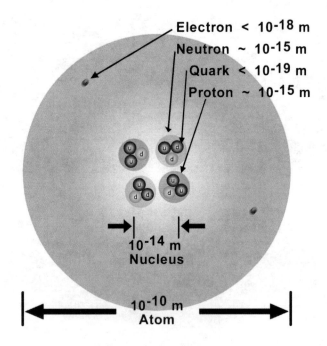

FIGURE 8.1B Zooming in to look at the inner workings of atoms, nuclei, protons, and neutrons with artist's conceptions and Feynman diagrams. Note that none of the pictures are drawn to scale and since quarks and electrons are both wave-like and particle-like, it is not possible to really draw their locations or the paths they take.

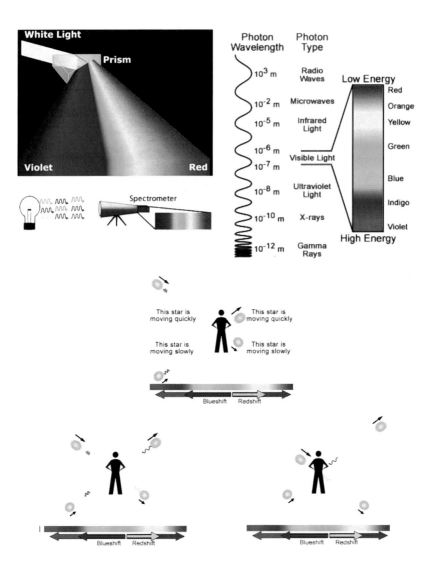

FIGURES 5.2b & c In "b," we see white light shining into a prism and separating into different colors; each color has its own wavelength. We now know that we can have light of any wavelength, and we give many of them names. While the visible spectrum is the only part we can see directly with our eyes, it is only a small part of the different types of light that exist.

FIGURE 5.4 TOP ROW An oversimplified view of a hypothetical star that emits only green light and is moving with a speed close to the speed of light. If we know the star emits green light, we can tell from the color of the light hitting us if it is coming toward us or moving away from us. We can also tell how fast it is moving. The three pictures show snapshots in time as the light from the star moves toward our observer. Note that the speeds are greatly exaggerated.

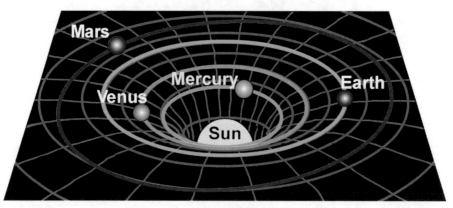

FIGURE 6.3 TOP ROW Two artist's conceptions of general relativity. Heavy masses stretch space-time and cause dents in space-time and other masses move in this curved space-time; an analogy is shown in the top image with pennies moving in a gravity well. With these ideas, general relativity describes the inner planets moving in the curved space-time around the Sun, as shown in the bottom.

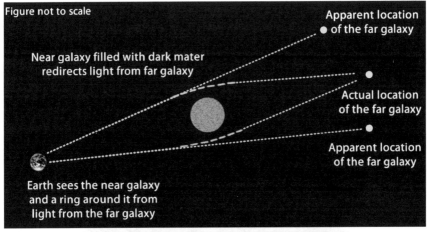

Figure not to scale

Apparent location of the far galaxy

Near galaxy filled with dark mater redirects light from far galaxy

Actual location of the far galaxy

Apparent location of the far galaxy

Earth sees the near galaxy and a ring around it from light from the far galaxy

NASA, ESA, A. Bolton (Harvard-Smithsonian CfA) and the SLACS Team

FIGURE 6.7 One way that light from a distant galaxy would look to us here on Earth when there is another galaxy directly in the way. In the special case that one galaxy is directly in front of another, there can be so much mass in the near galaxy that its gravity lenses the light from the galaxy behind it and creates what is known as an Einstein ring. An example with real data is shown in the bottom.

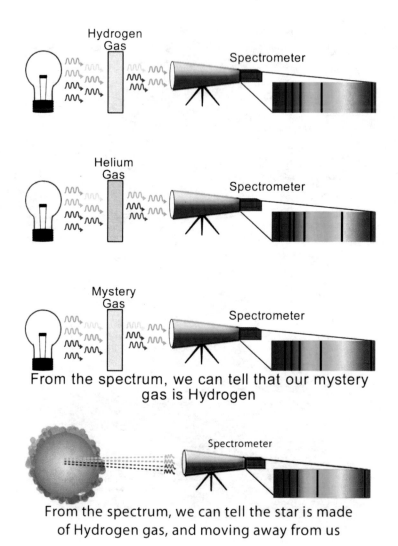

Hydrogen Gas

Spectrometer

Helium Gas

Spectrometer

Mystery Gas

Spectrometer

From the spectrum, we can tell that our mystery gas is Hydrogen

Spectrometer

From the spectrum, we can tell the star is made of Hydrogen gas, and moving away from us

FIGURES 8.2 AND 8.6 Different types of atoms have different energy levels. As the white light shines on the atoms, the electrons move between these levels and absorb only special colors. Since the energy levels are so distinct, this allows us to "fingerprint" the different types of atoms. The bottom figure shows a simplified view of looking at light from a star. By looking at the light from stars using a spectrometer, we can tell what they are made of and how they move. Note that the spectral lines are shifted to the right (red-shifted).

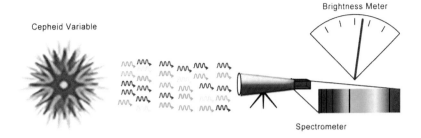

Cepheid Variable

Brightness Meter

Spectrometer

FIGURE 10.1 An artist's conception of one of the techniques to measure both distances to galaxies and their speeds and directions. Here, our simple brightness meter is counting the number of photons observed in the spectrometer. By looking at many stars in a galaxy, we can study the amount of light from them—as well as their Doppler shifts—to see whether the galaxy is moving toward us (blue-shifted) or away from us (red-shifted).

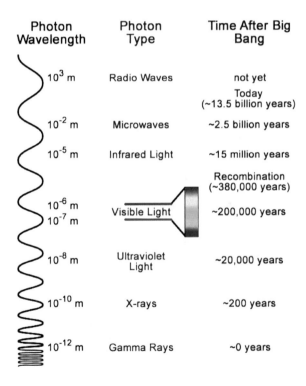

Photon Wavelength	Photon Type	Time After Big Bang
10^3 m	Radio Waves	not yet
		Today (~13.5 billion years)
10^{-2} m	Microwaves	~2.5 billion years
10^{-5} m	Infrared Light	~15 million years
		Recombination (~380,000 years)
10^{-6} m / 10^{-7} m	Visible Light	~200,000 years
10^{-8} m	Ultraviolet Light	~20,000 years
10^{-10} m	X-rays	~200 years
10^{-12} m	Gamma Rays	~0 years

FIGURE 12.7 A photon's wavelength stretches as the universe expands. The photons in the early universe were high-energy gamma rays. Now they are low-energy microwaves.

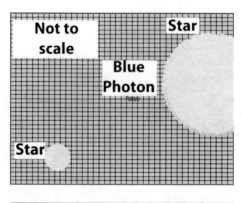

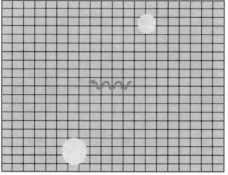

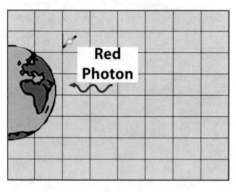

FIGURE 11.2 A photon in expanding space-time. We see light from a star in a distant galaxy on its path toward the Earth. The photon was emitted from the star at the top as a blue photon. In the middle plot, we see the photon as it travels through space and past other galaxies; it has traveled for so long already that the expansion of space-time has significantly stretched its wavelength. At even later times, the photon arrives at our eyes on the Earth having been stretched into a red photon. This stretching of the wavelength of light in an expanding universe explains the red-shifts of light from distant galaxies. It will also help explain the cosmic background radiation in Chapter 12.
Note that the grid marks are not to scale.

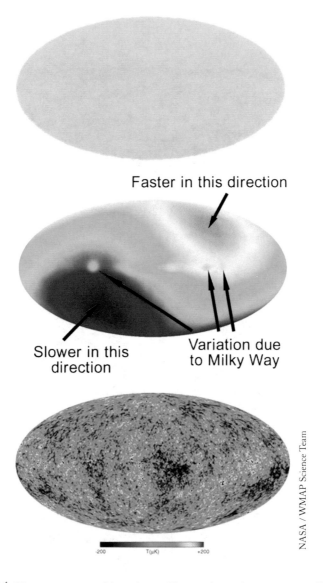

FIGURE 15.4 The temperature of the universe. The top shows the temperature distribution of the universe in terms of color in all directions. The middle shows the temperature
distribution after we subtract the same amount in every direction. The big variations (color
differences) are due to Doppler shifts because we are moving relative to the cosmic background radiation; the little variations in the middle are from the Milky Way's light. We can
subtract off both of these effects, as well. The results are shown on the bottom plot. These
small temperature fluctuations in the cosmic background radiation give us confidence that
our model of when galaxies form is correct.

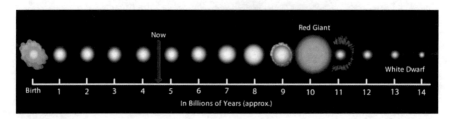

FIGURE 16.4 The life cycle of our Sun, which is a typical star of medium mass. The Sun is about 4.5 billion years old

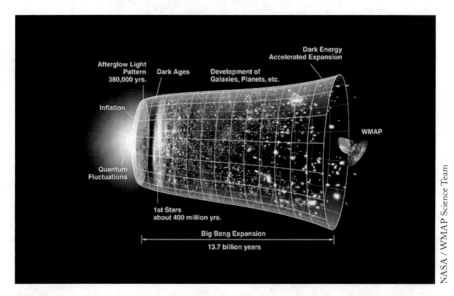

NASA / WMAP Science Team

FIGURE 18.4 A view of the evolution of the universe. Our best understanding today is that the universe will expand forever.

BIG BANG, BLACK HOLES, NO MATH

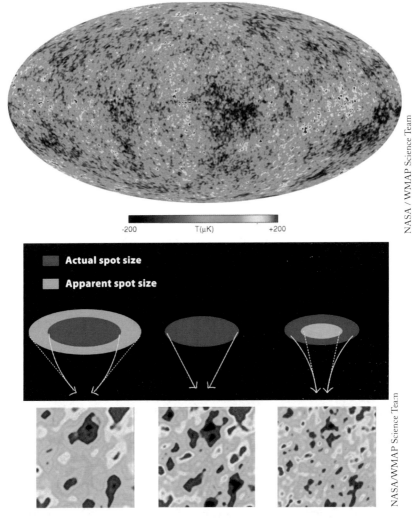

NASA / WMAP Science Team

NASA/WMAP Science Team

FIGURE 18.3 The top plot shows temperature fluctuations in the cosmic background radiation. These tiny fluctuations in the measured temperature distribution appear in different directions in the universe. The middle row shows how light from high (or low) temperature spots travel through the curved space-time of the universe, and reach us today. On the middle-left is a simulation of what would happen to the light from a spot as it travels through a universe that will end in a big crunch. On the middle-right is one that will expand forever. Between them is what would happen to light in a universe with a critical density (overall zero curvature of the universe). Since we know what the true width of a spot should be, we can use computer simulations to do a full prediction of the cosmic background radiation in the three different density scenarios. This middle simulation most closely resembles the data. To get a sense of the size of the spots in the middle plot, consider the size of the Moon as seen from the Earth. That size would cover about the diameter of a spot.

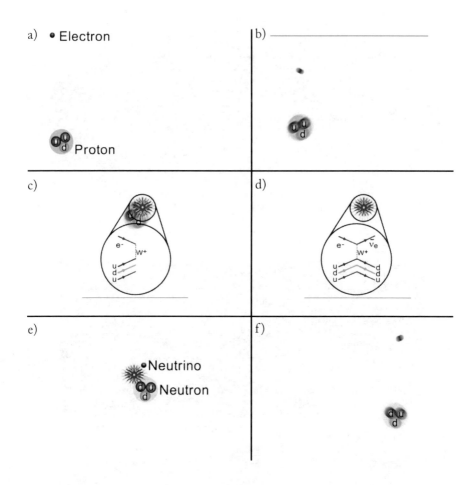

FIGURE 19.2 Weak interactions that were important in the early universe, as well as in the formation of neutron stars. The set of shots illustrates the joining of an electron and a proton to create a neutron, as occurs in the formation of neutron stars. Note the Feynman diagram of the interactions in the insets.

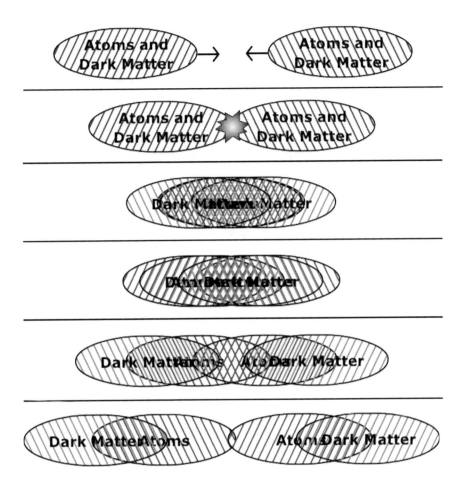

FIGURE 19.5 A series of action shots showing clusters of galaxies made up of atoms and dark matter on a collision course. Each has a distinctive way of interacting when it collides. The atoms will move through each other slowly because of their interactions and the dark matter will pass through quickly. During the course of the collisions, the two parts of the cluster will separate. As in Figure 19.4, the dark matter figures are our best guess as to what will happen with dark matter because we do not yet understand its true nature.

X-ray: NASA/CXC/CfA/M. Markevitch et al.; Optical: NASA/STScI; Magellan/U.Arizona/D.Clowe et al.; Lensing Map: NASA/STScI; ESO WFI; Magellan/U.Arizona/D.Clowe et al.

FIGURE 19.6 Real data showing a study of two colliding clusters of galaxies. The image's blue region shows where the dark matter is (using gravitational lensing) and the red region shows where the atoms are (using light from the collisions). This data is interpreted as two colliding clusters of galaxies, each of which started as a collection of atoms and dark matter. This set of observations could be explained by the pictures in Figure 19.5.

BIG BANG, BLACK HOLES, NO MATH

Fermi National Accelerator Laboratory, Office of Science/U.S. Department of Energy

Lawrence Livermore National Security, U.S. Department of Energy

FIGURE 19.8 Aerial views of the Tevatron particle accelerator at Fermilab (top left, four miles around the bigger of the two circles in the picture) and the Large Hadron Collider (LHC) at CERN (bottom, about seventeen miles around the circle). Each has created many high-energy collisions that might have produced dark matter particles. Scientists have studied these collisions to both look for dark matter and test supersymmetry.

FIGURE 17.5 A stellar black hole (left side of the drawing), which can be identified as it rips off atoms from a nearby star (right side of the drawing). As the material falls into the black hole, it produces the two beams of light we can see.

X-ray: NASA/CXC; Optical: Digitized Sky Survey